TEMPORAL INTELLIGENCE (TQ)

for Leadership and Organizational Development

Beyond EQ and IQ

WRITTEN BY

Flynn Bucy Ph.D. and Rich Doherty

Published by EDK Books and
Distributed by EDK Distribution, LLC
edkbooksanddistribution.com
edkbooksanddistribution@gmail.com | (206) 227-8179

Temporal Intelligence (TQ) for Leadership and Organizational Development - Beyond EQ and IQ

ISBN: 978-1-956065-14-5

Library of Congress Control Number: 2023917540

Printed in the United States of America

Illustrator: Mathew Miller

Editor: Barbara Kindness

Cover Design-Book Layout: Julie K. Lee @ Lee Creative

"We spend our lives on what we pay attention to"

DEDICATION

To Elliott Jaques and his decades of research
on how time is a critical element to
high-performance leaders and organizations

CONTENTS

PART TWO – *Time as a Frame of Reference*

PART THREE – *Cycle Timing*

INTRODUCTION

What do you believe is the most underdeveloped area of your leadership capacity? Planning? Delegating? Motivating? Communicating? There is a deeper capacity or intelligence that shows up in all of these core leadership functions—**temporal intelligence.** Being smart about the many ways we use the different dimensions of *time* is temporal intelligence.

Temporal intelligence is essential for good leadership. That is the thesis of this book. Unfortunately, most people have never heard of, or even thought about, how important temporal intelligence is to leadership and organizational effectiveness. Temporal skills refer to the range of awareness and capacities for understanding and utilizing the multiple dimensions of the term *"time"*.

Temporal intelligence is essential for good leadership.

There are multiple ways we use the term "time" to describe different temporal dimensions. Most leaders have an intuitive sense of timing, as well as time-use patterns, and time frames; but few have a conscious foundation in these dimensions of temporal intelligence. And fewer still have intentionally developed practices that can dramatically improve their temporal skills.

This book is about discovering, refining, and leveraging this important leadership asset. Temporal intelligence is more than simply managing your time well. It is understanding that different time dimensions are essential assets that apply in a wide variety of organizational and leadership situations.

Increasing temporal intelligence is to increase your understanding of how you and others construct, manipulate, and utilize the various dimensions of time. To discover this capacity is to uncover a valuable resource that is very often underdeveloped and underused. In fact, addressing temporal intelligence is a unique angle on how to improve your personal leadership and the performance of those around you. So temporal intelligence, or TQ, refers to a person's *awareness* of the multiple dimensions of time and *capacity to use* those insights in personal and professional practices.

The term "TQ" is an umbrella label used throughout this book to cover a range of concepts and capacities related to different ways the term "time" is used. However, the centralizing concept of temporal intelligence is that we can learn to focus conscious awareness on how we use the different dimensions of time and develop greater agility in how we leverage this valuable asset in our leadership roles.

Sometimes, it is also helpful to point out what something is ***not.*** In this case:

- **TQ is not** traditional time management. Although we discuss Time-as-a-Resource, the focus is not on techniques to be more efficient or to manage one's own calendar, Inbox, and to-do list.

- **TQ is not** learning new content about the enigma of time – it is not a study of time as a subject but is a set of concepts and practices about how to use temporal skills to be a more effective leader.

- **TQ is not** a single "thing." TQ includes a range of ways we can understand and use different dimensions of how we use the term "time".

- **TQ is not** simply a set of techniques. Practical techniques are a part of this book, but the foundation of TQ is based upon a person's ability to shift their perspective to gain greater awareness of temporal dimensions and greater agility to adapt time-use patterns.

Everyday language also demonstrates how we use the term in different ways. For example, some common phrases include:

- What time is it?

- How much time do we need to get this project finished?

- What is the long-term plan?

- Did you have a good time?
- Once upon a time…
- How have times changed over the last few decades?
- What is the time-space continuum?
- How much are you paid for your time?
- What do you hope to accomplish in your lifetime?

Taking on a concept as broad and integral as our temporal dimensions in a practical way is a challenge. The enigma of "time" has been the topic of books and conversations for thousands of years. One of the reasons the concept of time continues to be an enigma is that we use the term in multiple, yet different, ways.

As an example, in their excellent book *Time and Human Interaction: Social Psychology of Tim*e, Joseph E. McGrath and Janice R. Kelly point out the challenge of dealing with concepts about "time". "There are potentially two hundred fifty-six different conceptions of time that have been identified. Given the plethora of time conceptions, time frames, and other time terms, it seems worthwhile to lay out a set of terms that provide a standard language for "speaking of time."

The purpose of this book is not to produce another book on the philosophy of time (although we do enjoy those), but to focus on a few important dimensions that have practical value for leaders and growing organizations. In this book, we will be focusing only on three different conceptions of time.

• • • • •

How is Temporal Intelligence (TQ) similar to and distinct from the concepts of Emotional Intelligence (EQ) and Intelligence Quotient (IQ)?

Have you heard of, or worked with, the concept of *emotional intelligence,* or "EQ"? Emotional intelligence can be defined as the ability to perceive, use, understand, manage, and handle emotions. People with high emotional intelligence can recognize their own emotions and those of others, use emotional information to guide thinking and behavior, discern between different feelings and label them appropriately, and adjust emotions to adapt to environments.

Although the term first appeared in 1964, it gained popularity in the 1995 best-selling book, *Emotional Intelligence: Why it can matter more than IQ*, by Daniel Goleman. Goleman defined emotional intelligence as a set of skills and characteristics that drive leadership performance. Some researchers suggest that emotional intelligence can be learned and strengthened, while others claim it is an inborn characteristic.

Various models have been developed to measure emotional intelligence. In 1987, British psychologist Dr. Keith Beasley first published the term Emotional Quotient (EQ) after the Intellectual Quotient (IQ). "Intelligence Quotient is a score derived from a set of standardized tests designed to assess human intelligence The abbreviation "IQ" was coined in 1912 by German psychologist William Stern.

There has been a long-standing debate concerning whether "intelligence" can be developed or whether it matures along a predictable maturation curve. We will come back to this when we discuss how our cognitive capacity affects our temporal capacities.

EQ recognizes that our emotional experiences dramatically affect our lives and work. Emotions are always present, but many of us are unaware of our emotional states and unable to manage our own or others' emotions. By bringing our emotions into conscious focus and using various perspectives and practices, emotion can be harnessed as a powerful personal and leadership capacity.

Temporal intelligence, or TQ, is similar to EQ in the sense that time is always present but often unconscious. The mechanical aspects of time are all around us in clocks, mobile devices, calendars, and computers, but the time itself is an invisible intangible. Like our emotions, we all structure it differently and experience it in ways that are unique to ourselves. TQ and EQ are similar in that they have dramatic impacts on relationships and organizations. The two concepts are also similar in that they are capacities that can be developed with intentional effort.

Temporal intelligence, or TQ, is similar to EQ in the sense that time, like emotion, is always present but often unconscious.

There are also significant differences between EQ and TQ. Even if we are often not conscious of our emotions, they are generally accessible. Everybody has experience in dealing with their own and others' emotions. But even though time is a central feature in how we understand reality, it is generally not addressed directly. Time is all around us; we're usually just not conscious of it. We are like a fish being unaware of water.

The three temporal dimensions of TQ addressed in this book

The term *time* can be confusing because it is commonly used in multiple ways. This book will focus on three different dimensions of time that are extremely important for leadership and organizational performance:

- Time as a Resource
- Time as a Frame of Reference
- Cycle Time

TIME AS A RESOURCE: "Time is money" is an adage that captures the point. Estimating the calendar time and the number of hours some project will take is based upon how we manage time as a critical resource. Time can relate to both level-of-effort and calendar units, but they always go together to get things done. Individual time management is a foundational leadership skill, but in broadening the scope of the term, the ability to manage the temporal resources of an organization is the ultimate time management challenge.

When we think of time as a resource, we are thinking about the utilitarian value of units of time—like hours, days, months. When viewed this way, temporal units are scarce and valuable. From this perspective, we are trying to get the most value out of our limited time. And we are trying to get our people organized to use their time well. The problem is that most of us are not very conscious of how we and our people actually use time and how we might become better "time resource" managers.

TIME AS A FRAME OF REFERENCE: "Time frames" refer to the temporal dimension of the frame of reference we use to understand and address any specific situation, problem, or opportunity. Time frames are based upon using units of linear time (i.e., hours, days, weeks, years) to create the temporal context, or container, within which problems are analyzed and decisions made.

Every situation, or problem, or opportunity, has a time frame. However, often the time frame being used to frame the issue is subconscious. We often assume that others are using the same time frames to understand the same problem or opportunity. This is very often not the case. Different people using different time frames of analysis causes a great deal of confusion and lack of alignment in organizations. The time frames establish the relevant temporal horizon for planning and action.

CYCLE TIME: Cycle time is focusing on how things work together over time to bring processes and cycles to completion. Cycle time reflects the life cycle of any activity or process. Understanding Cycle time is based upon recognizing that the "timing of things" is a dynamic process and not just a static measuring convention for units of time. Timing involves the synthesis of factors used to anticipate the occurrence of events or outcomes in the future. It is judgment about time: later rather than now, this before that. Timing determines "when" we think something will happen; good timing is when we get it right.

Everyone knows "timing" is very important, but we often have a hard time explaining what it is or why it is so critical. The ability to estimate when a cycle will complete, or when an outcome will occur, is a powerful leadership tool. Timing provides an estimate of the number of temporal units that some process or cycle will take. Good leaders naturally have good timing, but this book provides a set of concepts and practices to help improve how timing is used and how we can develop better timing.

This is who we are

It is always interesting to understand an author's background and credibility to speak with authority on a chosen subject. In this case, we—Flynn Bucy and Rich Doherty—are the authors of this book and we bring different but complementary experience to the ideas discussed in this book. Both of us have long, deep experience as consultants and coaches of leaders in large-scale organizations.

Our collaboration on this book is based upon the results we have seen in using time intelligence as a key element to solve real problems for real clients. We began working together several years ago around the organizational theories of Elliott Jaques, a noted Canadian psychoanalyst, social scientist, and management consultant. We both found his work centered around time and organizational development extremely valuable when dealing with clients. We hope to demonstrate that what seems abstract at first can be made quite practical as a leadership and organizational development tool set.

For nearly twenty years, I, Rich, had been using the theories of Elliott Jaques to diagnose issues confronting the executives I coached and the organizations that they led. Using Jaques's time-based diagnostic approach, I was able to help executives make rapid progress in their overall performance, but for many years I regarded this approach as purely diagnostic; a way to assess executive performance relative to time intelligence and a way to assess that performance relative to organizational structure and design.

I had not developed an explicit method to incorporate time intelligence into my coaching work with clients. But, working with senior executives over extended time frames gave me a deeper appreciation for the role that time intelligence played in their success. I had the opportunity to observe leaders pursue strategies that unfolded over a decade or longer. It was apparent that they applied a distinctive capability to manage complexity over longer time frames.

This was the outcome predicted by Elliott Jaques's theory. This experience, plus developing my own time intelligence, convinced me that I had an obligation to develop a deeper coaching method to help leaders more consistently experience this capability.

It was my interest in developing such a coaching methodology based on the theories of Elliott Jaques that led me to reach out to Flynn. I noticed that Flynn was cited in Elliott's last book, so I reached out to explore our common interest in Elliott Jaques. I discovered that Flynn had been applying Jaques's theories to his organizational consulting work with large and small organizations, including businesses, nonprofits, and government agencies. He had developed a time intelligence toolkit and consulting method that enabled his clients to address complex problems over longer time frames. It was soon clear that working together we could develop practical tools to coach leaders to become more "intelligent" about the different dimensions of time.

I, Flynn, had the great pleasure and privilege to have known and worked with Elliott Jaques during and after my graduate studies at George Washington University in the late '80s. We spent a lot of time discussing, even arguing about the nature of our mental processes and models. He had well-documented evidence that one's mental processing patterns limited the time-horizon within which one could envision a coherent, dynamic future. What we discussed and worked on together was to explore not just an individual's maximum time-horizon and mental complexity patterns, but how agile they were in intentionally shifting to different time frames and complexity levels.

I got a nice shout-out in his last book, *Human Capability:*

> *"A major step forward was taken by Flynn Bucy in a doctoral research study carried out under the supervision of Dr. J. Harvey at George Washington University. In this study, Bucy was able to demonstrate not only that Lawrence Kohlberg's studies of moral value used in problem-solving referred really to categories of complexity of information process, but also that these categories corresponded to our preliminary formulation of four work categories. In the course of this study, Bucy refined our formulations, and his work led directly to further refinement of the four types of mental processing used in our research."*

We will be using the first person to talk about our examples of TQ throughout the book. In many cases, we have had similar experiences and we decided it was easier for the reader to not try to keep track of which one of us is telling a personal experience. So "we" will tell our stories as "I".

How does TQ relate to both Leadership and Organizational Development and the work of Elliott Jaques?

As mentioned earlier, TQ is an umbrella framework for how to harness a different awareness of time. This can apply on a personal level or on a professional level. Although the personal implications and applications of TQ are superpowers, this book is focused on applying TQ concepts and practices to the professional level.

Organizational leaders are this book's target audience. Excellent leaders not only have well-developed skills in core capacities such as planning, delegating, evaluating, and communicating, but they also must design and develop high-performance organizations. This is the area of experience and expertise both of us have.

The challenge for leaders arises when different business functions or business units use these three time-denominated ways of understanding the different time-use patterns at differing levels of an organization using different time frames. In businesses or situations where everyone uses time and human resources in the roughly same manner, it is easy to identify and communicate about time as a resource. When there are differences, it commonly leads to miscommunication and frustration over getting work done in a particular time frame. The question, "Why are they taking so long?" often signals this misunderstanding. Embracing "cycle time" is a key to understanding this common problem.

Again, we have found that the work of Elliott Jaques provides a set of unique insights that can help individuals and organizations be more effective. Jaques spent his life exploring how people should organize themselves to get the best outcomes for all of its stakeholders: the organization, the employees, and society-at-large. Over his decades of research and consulting, he concluded that chronological time provided a unique objective function for evaluating the working capacity of individuals and the effectiveness of organizational structures.

These are two simple, but powerful, concepts. Time measures the capacity of an individual to manage *complexity* in their work, and human hierarchies have levels organized around different *kinds* of work. For example, frontline workers begin with the capacity to manage tasks that can be completed in a day or two. CEOs of large managerial hierarchies have to manage tasks that might take a decade or longer to realize. The layers in a well-designed hierarchy, or "Requisite" in Jaques's terminology, sort themselves into levels, or "Strata," reflecting the different time horizons of responsibility.

Although Jaques has been recognized as a respected and high-profile organizational theorist for decades, his work is generally not well-understood or applied. Some have interpreted the implications of his work as elitist or deterministic, such as "CEOs deserve to be CEOs and frontline workers deserve to be on the frontline." Nothing could be further from Jaques's intent. His research showed that the capacity to handle longer-term and more complex work had little or no correlation with gender, education, race, or ethnicity. Jaques's model on the development curves for individual cognitive capacity will be discussed in Chapter 4.

Jaques's research also showed that we develop along predictable trajectories as we mature. We all grow up along predictable "curves." His research showed that everyone has the capacity to develop in this regard. In fact, you are on a development curve right now. By building your conscious awareness of time and complexity, this book will help you develop faster and with more intention. We will be using Dr. Jaques's work extensively throughout the section in each chapter on Organizational Development Implications.

• • • • •

How is this book organized?

The three parts of the book will cover the three different temporal dimensions to be discussed: Part One—Time as a Resource. Part Two—Time as a Frame of Reference. Part Three—Cycle Timing. Within each part are chapters that identify different TQ aspects of that temporal dimension. Each chapter includes a brief discussion of key **TQ concepts**. There are sections in every chapter that address a relevant **leadership development practice** and one which explores the **organizational development implications** of that aspect of TQ under discussion. In addition, each chapter has a set of **reflections and actions** that may be helpful in developing greater TQ.

Throughout the book, we will be sharing a number of relevant examples and scenarios of our work as executive coaches and organizational consultants. The overall thesis of this book is captured in a broad generalization made by a current client, Ben Jacobson, who really embraced the new TQ skill set we had been discussing.

When asked how a new focus on TQ had affected his performance, Ben said, *"I had never really paid attention to time as a set of concepts and skills. I was motivated to just create the intention to be more aware of the different ways I use time and to see how many places temporal skills come into play in my leadership role. Increasing TQ does not take a lot of effort but requires a simple, and intentional, shift in perspective that creates new insights into the challenging problems I deal with on a daily basis."*

What is the value proposition for this book?

Our hope is that you will discover a new treasure trove of leadership skills and organizational development practices around TQ concepts. There can be almost immediate changes in some of your current practices. In addition, these skills and techniques can be refined over time. You will certainly find a lot of new and interesting ideas, but the approach presented is more than a handful of "good ideas."

The value proposition for this book also has three elements:

- Learning key concepts and practices can lead to greater TQ and more capacity to manage time as a resource, time frames, and cycle timing.
- Learning techniques and exercises can help develop TQ awareness and skills.
- Integrating TQ development into your leadership practice and organizational design improves performance.

After years of working with these concepts, it is hard for us to think of leadership situations that do not involve some dimension of time. One recent client summed it up by saying, *"TQ has not only made me a more effective leader, but it has also made my senior team and the whole company more effective in both the short term and the long term."*

"TQ has not only made me a more effective leader, but it has also made my senior team and the whole company more effective in both the short term and the long term."

Before we launch into the first part on Time as a Resource, we encourage you to work with this material to enhance your TQ skills and not just skim through for a few new ideas. Personal and professional development takes time, but it is time intelligently invested. The timing is right. So, let's get started.

PART ONE

Time as a Resource

INTRODUCTION

How often have you heard the expression "time is money"? In general, people nod their heads in agreement when they hear that phrase. How is time like money?

When was the last time you heard yourself saying, "This is a waste of time"? It's common for all of us to feel like there's not enough time, or that we can save time. What kind of "time" is this referring to?

This section introduces a new way of understanding and using the term *time* that includes the concepts of time as a commodity with a utility value.

Time understood as a resource is what the Greeks referred to as ***Telos* Time,** or the utilitarian value of time. In modern culture, time-as-a-resource is probably the most common way we understand and use the term "time". When time is thought of as a valuable commodity, there is a built-in scarcity. Much of modern finance is organized around increasingly sophisticated ways to exploit the element of time. Options and arbitrage are two simple examples of this. In one sense, time is our most valuable personal asset; the essence of what makes up a lifetime. But in another way, time is the raw material of organizational performance.

Time is the raw material of organizational performance.

Organizations are essentially based upon leveraging the value of work from multiple individuals around interdependent business processes. Organizations buy time from their employees and spend it to create value. Business functions require certain levels of human resource, and we denominate that resource into time units.

The resource of time is made up of both "calendar time" and "units of work." Business processes are designed assuming a specific amount of calendar time and human resource time are available to accomplish the required outcome. Telos Time includes the units of time invested by human resources, which have a cost, and a perceived value. The challenge is to improve the way we calculate the cost/value of the human resource time invested in getting work done—short-, medium-, and long term.

Time use is an expression of value and priority. There is limited time and how it is used reflects its perceived value. Level of effort over a specific time frame determines the cost in human resources necessary to accomplish a specific outcome.

Leadership is largely a matter of managing the delegation of work units of differing skill sets to resource multiple, parallel action plans. Becoming more aware of how our time and the time of others is spent, or invested, is an essential TQ skill. Awareness of time use creates the opportunity to refine our capacity to forecast time resource requirements and to understand the time use requirements of other individuals and functions. "How can we help?" replaces "Why is it taking so long?"

TQ Capacity Development for High-Performance Leaders and Organizations

Jaques's first research breakthrough was recognizing the correlation of task-time, the essence of time as a resource, with equitable pay. In the 1950s, he was employed as a consultant by Wilfred Brown, the CEO of the Glacier Metal Company. One of his assignments was to find a way to create pay parity across different functions and levels of work.

The question he was trying to answer was, "What principle would lead both an accounting clerk and a lathe operator feeling they were paid a fair wage?" What he discovered in that initial study was a theory that informed his work throughout the rest of his career. The time span of control experienced by any worker in any function, whether it was an hour or a year or longer, determined what they considered fair pay, regardless of their role or function. This finding was validated numerous times over the following sixty years by Jaques and other researchers.

The significance for our **Time as a Resource** lessons is the basic understanding that everyone experiences the time as a resource utility as an aspect of their work. We all feel time; the "weight of responsibility". The longer the time duration of a task, the more we feel the time pressure. Even in those periods when our work progresses in an orderly manner with few time constraints or conflicts, we're feeling the time element of our longest task. Tuning into this personal experience of time is one of the first steps to appreciating the dimension of time as a resource for ourselves and those around us.

• • • • •

Case Study

The case study in this section is based upon coaching and consulting work with a CIO at a major retail food producer and distributor. One of the constant challenges he faced was balancing human resources requirements and costs with pressure to do more-with-less, and faster. As we will discuss in this section, the root of his short-, medium-, and long-term problems was embedded in the different ways different senior leaders understood the calculus of Time as a Resource.

My client was Richard Smoot. I had worked with Richard previously and understood his two sides. On the one hand, he was quite intelligent and had a good sense of technology, people, and politics, but on the other hand, he was often extremely hard-headed and would not change his mind or expectations when "things changed"—as they always do.

The acute problem that was giving him enough pain that he called me in was driven by a budget-cut mandate, while at the same time, he was falling behind on key projects and racking up a huge "legacy debt" by pushing off important but not urgent investments in development, infrastructure, and security.

The company had recently been purchased and had gone from one of the largest family-owned companies to just one of the many companies under a new umbrella. This brought a whole new set of challenges.

Expectations were out of touch with reality. Evidently, promises had been made during the transaction negotiations that created unrealistic time frames for converting to a whole new system and releasing a set of new "user-friendly apps". In addition, there was a mandate to cut costs.

To compound the problem, different senior leaders had different priorities with different deadlines and they were usually not clear of other executives' expectations—creating what seemed to be an impossible situation. So, Richard brought in a consultant—me. Over time, we discussed a set of key concepts and dug into different aspects of his dilemma.

We will use scenarios about Richard and his challenges throughout the three chapters in this part.

• • • • •

Overview

Chapters in this first part use the same framework as the first two parts. There are key concepts, key practices, and learning exercises in each chapter.

Chapter 1 – *Managing Time as a Resource* is based upon the recognition that there is limited time and how time is used is a value statement. In addition, a central thesis of this chapter is that time is not a "thing" but the measure of a dynamic—the dynamic of constant change. Managing time is managing change.

Chapter 2 – ***Awareness of Time-Use Patterns*** explores how we can improve our time use by being more aware of our time-use patterns and be more intentional about how we spend our own time in the future. In addition, there is also great value in understanding how others in your organization are spending their valuable time. Agreeing on time-use is agreeing on priorities.

Chapter 3 – ***Estimating Time Requirements*** addresses our internal calculations for how much of the resource of time will different activities require. Insight into how we use mental algorithms to estimate time requirements provides a basis for improving our TQ skills.

This part of the book addresses new TQ concepts about time as a resource, but the real value to you as a leader will be determined by how you work with these concepts to improve your actual awareness of time-use patterns and your ability to accurately estimate how many human resources a project needs.

CHAPTER ONE

Managing Time as a Resource

How often do you feel like there is just not enough time in a day? When was the last time you felt overwhelmed by too many balls in the air and not enough hands to juggle them? There seems to be a crisis of time scarcity in our busy lives. How can we learn to better manage the flow of action items, expectations, and responsibilities—personal and professional?

Stress, frustration, and despair are common emotional signs of poor TQ. The problem starts with the fact that our time is limited, and our potential uses of time are infinite. Learning to better manage time as a resource is a valuable skill set, but it is based on reframing time-use to manage the *flow* of time instead of trying to control the *rate of change*.

The term "Time Management" has developed a bad reputation. Most books and consultants focus on getting more done with less time or trying to become more efficient. This book is about trying to *work* with time instead of *control* it. But TQ is about being smart about time. It is about effectiveness, not efficiency. TQ reframes time management to focus on managing complexity and not trying to solve one problem at a time.

The reality is that we spread our time-use among different dimensions of our lives. For simplicity's sake, let's divide how we spend time on our personal and professional agendas. In reality this is a false dichotomy, but it keeps our focus on TQ for leadership in an organizational setting. Leadership is largely about managing the time-use of others to accomplish a range of short-, medium-, and long-term objectives.

Thesis: Managing organizational time-use patterns presents a different set of challenges than simply focusing on one's own schedule. The level of effort over a specific time frame determines the cost in human resources necessary to accomplish specific results. This chapter looks at three levels: managing your own time, managing the time-use patterns of your direct reports, and managing the human resource time requirements.

This chapter is based upon the recognition that there is limited time and how time is used is a value statement. In addition, a central thesis of this whole section on Time as a Resource is that time is not a "thing" but the measure of a dynamic—the dynamic of constant change. Managing time is managing change

---

How time is used is a value statement.

TQ skills are based upon developing conscious awareness of the flow of time and improving our ability to manage the complexity of rapid change.

Overview of TQ Concepts:

- Time management for humans
- The Efficiency Trap
- Time use is an expression of value and priority
- Managing time is managing change
- Spending time vs. Investing time

Leadership Practice: Time chart of accounts and budget

Case Study Scenario

Richard began our first face-to-face consulting meeting with a blunt, but misguided, and apparently unanswerable, question:

"How do we get more out of less?"

He described a set of critical projects that needed more resources if they were going to be delivered on time, and a mandate that he was going to have to cut his budget across the board. He was desperate.

"How can I get out of this trap?" he asked, with a tone that suggested I had an answer.

"I do not know how to do what can't be done," I replied. "But let's start with a few basic features of reality. There is a direct relationship between your productivity level and your access to resources. Your first problem is that your senior leaders are fantasizing, and they know it. They have made a bunch of promises and now hope you can solve their problem. Don't let them make *their* problem *your* problem."

"Go on," he said, as he sat up, paying close attention.

"Second," I continued, "the key is to establish the value proposition for different investment scenarios and let them deal with the trade offs, instead of expecting you to do the impossible. Nine times out of ten, you can clearly communicate the risks and benefits of what you know is the right long-term strategy. Let's talk about that."

TQ Concepts

Time management for humans

Time is the great equalizer. Everybody gets the same amount each day and chooses how to use it. In addition, as mortals we all have a limited time to be alive. I read an excellent book by Oliver Burkeman, *Four Thousand Weeks: Time Management for Mortals*. The title and thesis emphasize the fact that the amount of time we have in our lifetimes is limited. He estimates that if a person lives to approximately eighty years old, they will have only four thousand weeks to work with. With simple math, we can calculate that we each get one hundred and sixty-eight hours per week, and everybody uses it all every week.

Since we each have a limited time in a lifetime and in a week, and for that matter in a day, we need to start by understanding how precious it is. In a way, it is life itself. But it is not a commodity. Different ways we use time have different value propositions. How we use time reflects both its value among current time-use trade offs and its perceived value for future outcomes.

The conundrum we mortals find ourselves in is that we are faced with infinite options for how to use time, but a starkly limited amount of time to spread among multiple important parts of our lives. Work/life balance is a common phrase when dealing with these inevitable trade offs. For our purposes in this book, we will focus on the "work" side of the agenda. But the dilemma is still the same—infinite demands and limited time. And to make matters worse, within our professional world, we have multiple roles and goals and these all have short- medium- and long-term expectations.

The Efficiency Trap

What is the efficiency trap and why is it so seductive? Efficiency is often held up as the highest value in our market-based worldview—getting more out of less. Many time management books focus strictly on this limited objective. In the short-term, productivity gains can be accomplished through more efficient use of inputs, but in the medium- and long-term, this can be a trap. The efficiency trap is a result of not coming to grips with the underlying reality that we cannot do it all. That we *must* do more than we *can* do does not make sense.

According to Burkeman, "If you truly don't have time for everything you want to do, or you feel you ought to do, or that others are expecting you to do, then you don't have enough time to do it all—no matter how grave the consequences of failing might prove to be. So technically, it's irrational to feel troubled by an overwhelming To-Do List. You'll do what you can, you won't do what you can't, and your inner voice and your outer boss are simply mistaken." Most of us don't think that logically.

Why is getting caught in the efficiency trap so common? It is seductive to think you can get more out of life if you manage your time like a transaction. Getting more for less. Most people in our culture only one want thing—more. It seems like our limited time is a constraint on doing more, having more, being more. And so, it feels like if we just had *more* time, we could get more of everything else.

The fact that we have a limited amount of time and an unlimited number of things we want more of leads to the realization that we cannot solve the problem by being more efficient. My point is: We have to reframe the issue to focus on *effectiveness*, instead of efficiency.

In my experience as an organizational consultant, I've seen again and again that a common mistake is to assign more to a person than they have the "bandwidth" to manage. It is like pouring a gallon bucket into a 16-ounce cup—so water is spilled.

The unstated assumption is that the person can cram in more by being more efficient or working harder. Unfortunately, what often happens is that the person "spills" whatever does not fit into the cup. There is a tendency to work on short-term tasks and "spilling" the longer-term tasks associated with their role. The question is, which part was allowed to spill and what parts were contained? Or was quality sacrificed along the way to squeeze in more? (Fig. 1a)

The expectation of doing more with less can have serious emotional consequences for individuals and significant performance consequences for organizations.

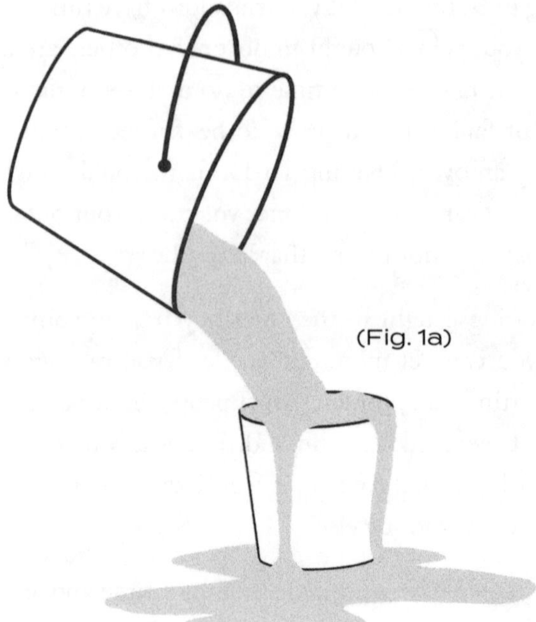

(Fig. 1a)

I have coached several new executives who experienced this problem, particularly after being promoted to the VP or General Management level. They struggle to balance the necessity to use their teams to address the near-term issues with the longer-term responsibilities that are the core of their jobs. When talking to their managers and peers, what I typically hear is, "I need her to be more strategic in her outlook. All she wants to talk about is stuff happening next week or next month. That's not her job."

My coaching to the manager of the new executive is to accept the reality of the transition and limit the new VP's workload to what is doable given their capacity and experience. Using that approach, the new VP can get up to speed on their actual responsibilities and not tread water on shorter-term responsibilities.

Microsoft co-founder Bill Gates observed on numerous occasions that, "Most people overestimate what they can do in one year and underestimate what they can do in ten years."

"Most people overestimate what they can do in one year and underestimate what they can do in ten years." – Bill Gates

Coaching new executives is frequently about helping them see the bigger picture for their work. The TQ alternative approach is to avoid the efficiency trap by focusing on longer-term value instead of short-term activity. We should be more intentional about what to leave out instead of trying to cram more in.

We should be more intentional about what to leave out instead of trying to cram more in.

Time-use an expression of value and priority

How often do you find yourself logically thinking through your time-use trade-offs? Most of the time we make decisions about time-use subconsciously, outside of our logical analysis, but every time-use choice is a trade-off. Any time you spend on one agenda means not spending that time on others. Every decision of where you spend your time is a statement of value and priority.

When Rich starts coaching a new executive client, he asks, "Show me your calendar?" He's not interested in how efficiently the client is using their time, he needs to understand their priorities, the trade-offs they make, and, most importantly, what they value.

Every time-use choice is a trade-off.

We are never fully aware of the value trade-offs involved when we prioritize. Our values are our own, and rarely fully aligned with the tasks, no matter how important, that are assigned to us. Everyone, from CEOs to new hires, brings unique values to the task of prioritizing their time.

Getting clarity of priorities can best be brought into conscious focus by stepping back and setting future time-use targets. When prioritizing how we intend to spread our time among conflicting alternatives, we need to be more consciously aware of how those targets reflect an endless, flowing array of short-, medium-, and long-term trade-offs.

TQ is developing our ability to dynamically prioritize between time-use intention and the changing demands of circumstances. There are no set right answers in terms of how to best use our time—it depends. It depends on how we value and prioritize the things we choose to focus our attention on.

One common problem I see with my clients is that they confuse "worrying" about something with "working" on something. In one case, a very senior executive was trying to explain how their time-use patterns reflected their priorities. He told me, *"At first, your assignment to think about how I would ideally spread my time among the seemingly endless demands on my time seemed straightforward. But when I really started looking at how I spread not just my time, but my focus and intention, I realized I was spending a lot of time worrying about problems I could not solve or circumstances I could not control. I was not really working on those issues. I was just spending precious time worrying about them. Your "time chart of accounts" exercise really helped me get this distinction into focus."*

Managing time is managing change

Time units are what we use to measure the rate of change itself. A day is one complete revolution of the earth. We're moving at roughly a thousand miles an hour at the surface of the earth to complete a single revolution. A year is one trip around the sun. We're moving through space at roughly 67,000 miles an hour to complete one annual orbit. That's a lot of change.

The perceived rate of change, or flow, is based upon our perspective—and time frame of analysis. Good leadership is about being able to manage the flow and dynamic of the many steps and stages required to accomplish any short- or long-term outcome.

When we objectify time as a "thing," our approach to managing units of time makes sense. However, time is also a dynamic that is measured in temporal units. Time is what is measured, as well as what we are measuring. No wonder the term can get confusing.

We confuse the units of time as a measurement convention with the dynamic flow that it is calibrating. Time units measure the rate of change we are focused upon. Like the well-known dilemma of physics concerning the question of whether light is a particle or a wave. This riddle is solved when we realize it is both—and either—depending on how it is measured. A particle is a snapshot or a unit of energy in time, while a wave is the flow of energy over time.

When I was in graduate school at George Washington University getting a Ph.D. in Organizational Behavior and Development, I took a course in "Change Management". This concept referred to the implementation of new systems or procedures. I have worked on numerous such organizational change projects through the years and have actually called myself a "Change Management Consultant" from time to time.

In our current turbulent environment, the term "time management" has come to mean something different. Instead of being thought of as integrating planned changes into current systems, change management is now used to mean adapting to unpredicted and uncontrollable changes occurring in markets, in technology, and in our global culture.

The increasing rate of change is driving a new approach to leadership, reframed to think in more dynamic ways in order to lead in more dynamic times. A simple example of this principle is a sailboat. Although the general destination is established, actually sailing the boat is a constant process of small adjustments based upon wind and weather.

Leadership excellence is about managing the dynamic among the parts of the organizational system and not getting caught up in the detail of the parts. The TQ approach focuses on managing complexity and not simply trying to solve specific problems.

The TQ approach focuses on managing complexity and not simply trying to solve specific problems.

Spending time vs. Investing time

If time is money, what do we spend it on? Or do we invest our time? Or both? It is important to distinguish between the cost of time as a consumable resource and the return on that investment. How do you distinguish between wasting time and saving time? When thinking of time as a resource, how can we best think about how we manage this valuable resource? Since we have a limited amount of time in a day, week, or life, we want to use it well. So how do we think about how we use time?

"We spend our lives on what we pay attention to"

Time-value calculations can be thought of as analogous to the three primary financial statements: cash flow, income statement, and balance sheet.

- Cash flow statement tracks cash coming in and going out over some period. In a similar way, a time sheet reflects where time is spent.
- An income statement reports the income and expenses over a period. In our time analogy, this is like assessing what was delivered for the time spent.

- When we use balance sheet logic, time is an asset that is invested in expectation of value over time and not simply a cost in the period. It is learning to think of the resource of time as an investment in longer-term outcomes instead of limiting the scope of analysis to any one reporting period.

One of the capabilities of elite leaders is to think longer-term and understand the return on investing human resources in the present with the expectation of high payoff over time. This can be thought of as Return on Time Invested, or ROTI, instead of ROI for financial investments.

There are also significant opportunity costs associated with how time is spent or invested. When time is committed to one priority, it inevitably takes away that resource from other tasks. It is always difficult to imagine what might have been down another fork in the strategic road, but leaders constantly battle with this trade-off. Time as a resource is limited and expensive. A person's "time-use portfolio" is one way to assess value.

Leadership Practice: Time chart of accounts and budget

Description:

This practice provides a foundation for structuring and prioritizing your time use. Developing a time chart of accounts and budget is similar to establishing a financial structure for where money is budgeted and where it is spent. This practice provides a framework for how you divide your time use into different categories. Think of it as where you would "bill your time" if you kept a time sheet.

Each time-use category should have an "account" where it is budgeted and tracked. In a similar way, a time chart of accounts is a logical listing of where you spend your time. This is a practice to identify your primary time-use categories, establish a "time budget," and begin tracking your actual time-use patterns. Once you have established a set of categories where you will "charge" your time, you can then "budget" or plan your time-use patterns to reflect your ideal balance.

Process Steps:

Step 1 – Develop a personal time chart of accounts reflecting the different parts of your job and life. When asked where we spend our time, to answer we need to have a framework, or classification scheme, to sort time-use into different categories.

Because this book is focused on TQ for work roles and not personal TQ, the focus here is categorizing the different parts of one's job. However, the same approach can be applied to *personal* time-use. In fact, in my own practice, I start with a comprehensive framework that includes all of my time-use categories. I use 1000 through 9000 to distinguish different time-use priorities. I use the 2000s for work time-use as described below.

Start with a longer-term perspective, such as, "in order to reach my professional goals over the next ten years, how should I spread my time?" Then think about how to balance time in the medium term—let's say ninety days. It helps to start out with how you spread your short-term time over a "common week". This is the purpose of a time chart of accounts.

An example of how a leader might categorize their work time-use:

100	Direct engagements on active projects (this can be divided among major projects)
200	New opportunity development
300	Research and analysis
400	Professional development
500	Administration and reporting
600	Special projects
700	Mentoring and team building
800	Community service
900	Other

Step 2 – Once you have established the framework, establish a "target level" for each category and major sub-category. Fill in your "time-use budget" or the aspirational balance among the competing categories.

Step 3 – Recognize that time demands vary from day-to-day and week-to-week, but the target should be based on the spread over a few months. Think through your targets within a quarterly time frame.

Step 4 – This is an ongoing step to adapt the target model and reflect changes in circumstances and priorities.

Once when I was working with a client to set up a Time Chart of Accounts and budget, he shared how he had spent some time reflecting on the practice. A personal analogy provided real insight: "I'm a big gambler," he said. "I love to go to Vegas. I particularly like playing slots." Then he asked me, "Do you know what you get when you win on a nickel slot machine?"

"No," I replied, not considering the obvious.

"Nickels. And what do you get when you win on a dollar slot machine?"

"Dollars?" I sheepishly asked.

"There you have it," he said. "I realized I had been playing the "nickel slots" with my time and not getting much back even when I did win. The things that are important are like dollar slots. What I discovered through your time-use analysis exercise was that I had been putting dollars into a nickel slot machine and expecting to be paid off in dollars. I needed to learn to put my time into the things that are more valuable. By that I mean putting my time into projects that would have the biggest payoff in the long run and not just cha-ching in the short term."

It will soon become apparent that these concepts can be put into practice in other areas of your life—not just your workday. In fact, it may become a very beneficial habit.

TQ Organizational Development Implications: Structure vs. Content of Mental Processes

A central concept in the work of Elliott Jaques is based upon his use of the term "mental processes." He makes the distinction between two types of mental processes. One set of mental processes deals with the "content" of what we know. These thinking processes are logical, content-based, and verbally articulable. He also discovered that they occurred in repeating sets of five levels, or a Quintave. It is like an octave, with eight notes that start and end with the same note. We will be exploring Jaques's Quintave Model of Cognitive Functioning in later chapters.

Jaques defines *Mental Processing* as a wide range of cognitive functions, including the mental processes by which one takes in information, organizes it, and uses it to make decisions. But, as he points out, there is an important distinction between the "what," the content of the information gathered and processed, and the "how," the structure we use to make sense of the information coming in.

In addition to the cognitive, verbal logic of the content we work with, Jaques made an important contribution by making accessible to testing another type of nonverbal mental processes that he referred to as the *structure* of how we think. Understanding the distinction between "content" of information and "structure" of how information is processed is a multi-million dollar takeaway. It is within these alternative subconscious structures that we get insights into how time awareness and time frame capacity play major roles in organizations.

In his book *Requisite Organization* Jaques describes the four levels of logical structure. A graphical representation appears on the next page (Fig. 1b)

Declarative processing*: a person argues his or her position by bringing forward a number of separate reasons for it. The reasons are separate in the sense that each is brought forward individually on its own and no connection is made with any of the other reasons. For example, "Here's one reason for my idea and here's another and another as well."*

***Cumulative processing**: a person argues his or her position by bringing together a number of different ideas, none of which is sufficient to make the case but taken together they do. This method of processing has a pulled-together, conjunctive quality.*

***Serial processing**: a person argues his or her position by constructing a line of thoughts that make up a sequence of reasons, each one of which leads to the next, thus creating a chain of linked reasons.*

***Parallel processing**: a person argues his or her position by examining a number of other possible positions as well, each arrived at by means of serial processing. The several lines of thought are held in parallel and can be linked with each other.*

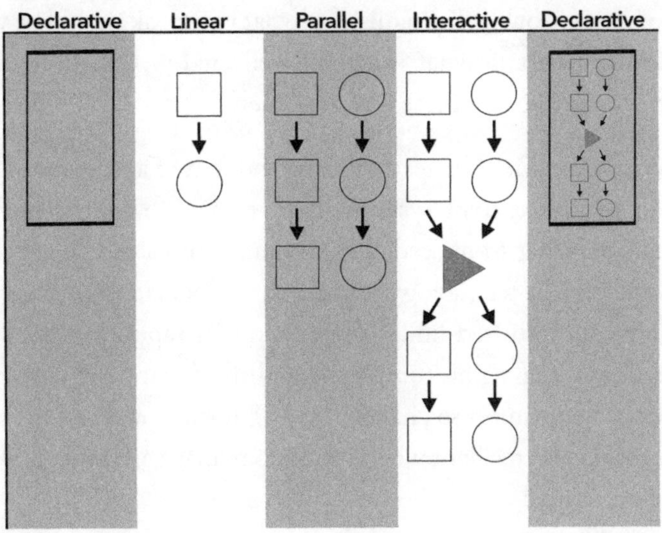

(Fig. 1b)

These different types of mental processing are not right or wrong. They simply reflect different levels of logical abstraction. Gaining insight into different people's processing style and level of complexity and abstraction helps make sure everybody is focused on their part of the system.

Reflections and Actions

- Take a moment and reflect on the stark fact that our time is limited. We have to make choices and trade-offs. Pinpoint a few areas where you are having a difficult time juggling priorities and time commitments. Are you just running faster to keep up with external demands or are you ready to stop, reflect and make some difficult, but important, trade-offs. Can you describe those trade-offs?

- Do you agree that how you spend your time is a value statement and is reflected in how you prioritize your time use? How does your time use reflect your values? People often confuse beliefs about their "held" values with the value actually represented in choices they make. Are you having trouble distinguishing what you think should be your value prioritization and what is actually reflected in your time use? What can you do to bring these into alignment?

- As quoted earlier, "Instead of being thought of as integrating planned changes into current systems, change management is now used to mean adapting to unpredicted and uncontrollable changes occurring in markets, in technology, and in our global culture." How has your role recently shifted under the pressure of rapid change? How does the idea of managing complexity and the dynamic of change fit with your current time-use challenges? What can you do to become more adaptable and resilient when trying to manage an unmanageable level of complexity?

- Think back on a few of your significant accomplishments. How long did you work on that project? How much time did you invest in reaching that accomplishment? Do you remember times when you felt the stress of taking time from other demands to invest in the eventual payoff? What are some of the bigger, longer-term projects you are working on now that you need to make sure you are taking regular time commitments to pursue? Have you thought through the ROI on those investments?

CHAPTER TWO

Awareness of Time-Use Patterns

How aware are you of where you spend your time? Do you often take time to reflect and consider how you are dividing your time among your range of interests? Do you think it is important to your leadership performance to be conscious and intentional about where you invest your time? My experience is that everybody thinks it's important but very few people pay attention to what they actually spend their time on. Most of us are pretty good about where we spent time today or this week, but we cannot accurately estimate where we spent our time over months and years.

Being aware of time-use patterns, your own and those of others in your charge, is a very important TQ capacity for effective leaders. But it is not an easy task. In this part we have been discussing time as a resource, but awareness of how valuable—and limited—that resource is will sharpen your focus on understanding *how* you spend your time. When elite leaders get clear about the value of increasing their awareness of how time is used, they make an intentional effort to improve their time-use patterns--their own and the people they are responsible for, and the unfolding of the projects they lead.

Thesis: When "time" is understood as a resource, awareness of how time is used is a valuable TQ skill. Like other TQ skills, becoming aware of how time is used requires a person to take a meta-perspective and intentionally notice time-use patterns; both as they are occurring and over time. Recognizing our time-use habits and awareness of the time-use patterns of others allows an individual to assess and realign how the resource of time is best invested for short-, medium-, and long-term outcomes.

TQ Concepts to be covered in this chapter:

- Awareness of our own time-use patterns
- Awareness of others' time-use patterns
- Time-use habits
- Actual time-use vs. budget of intentional time

Leadership Practice: Comparing actual time-use to target time-use

• • • • •

Case Study Scenario

Picking up on the Case Study with Richard for this part, let's focus on a particular scenario we worked with. Our first conversation had set a tone for a series of discussions we had over several months. Richard's focus was on specific new budget constraints while expectations were expanding.

In our previous interview, he asked, "How do I get my CEO and CFO to understand the value of the investments we need to make in our IT systems?"

When I asked for clarification, he said, "There just is not enough time in a month."

I replied, "Time is the great equalizer. We all have exactly the same amount of time: 168 hours in our seven-day week. . There is no more time.

"It is all about how you use both the calendar time and human resource time to meet priority outcomes. How clear are you on where your people spend their time? For that matter, how aware are you of how you spend *your* time? It sounds to me like our first step is to distinguish between different types of time and different types of time management."

Obviously confused, Richard looked back at me with a blank face and said, "What do you mean by different types of time? This does not sound very practical."

"On the contrary, understanding time as a resource underlies many of your most important decisions. Let's discuss how you categorize, monitor, estimate, and use time as a resource."

Again, with some exacerbation, he asked, "How is this going to solve my budget crisis?"

"Let me first provide some key concepts and practices and then we can explore how these tools can be used to address the specifics of your current budget crunch. And, to look for ways to avoid this kind of situation in the future."

"Well, I still have some questions," he responded.

"Great, we'll get to them soon," I replied.

TQ Concepts

Awareness of our own time-use patterns

How aware are you of your current time-use patterns? Can you give a reasonable estimate about how much time you spend on different parts of your job, and different parts of your life? Most people can't, but that is like not knowing where you spend your money. Time as a resource has value and being aware of how it is spent is critical for high-performance leaders.

You may have never thought about this before, but your time is your most valuable asset. How you spread that asset over multiple priority categories determines how effective you will be in your leadership role. Before you can have time management, you must develop time-use awareness and prioritization.

Time is your most valuable asset.

Awareness of your time-use includes recognizing patterns in multiple nested cycles. We have daily time-use patterns. We all spend 24 hours every day doing something. We are also spending time within weeks, months, years, decades. Where have you spent your time to date in those various time frames? And how do you want to distribute your time-use over those different cycles in the future? In other words, how will you spend your life?

As discussed earlier, shifting our awareness to notice specific features of our experience can be accomplished by shifting to a meta-perspective, stepping back from our stream-of-consciousness, and directing our focus to a specific issue or trend. Breaking into our conscious attention is required to be aware of our time-use patterns. The key is to set your intention to notice time-use patterns. This can best be done by taking two to three minutes each day to assess where you spent your time that day and do the same for the other cycles in your life. Assess the cycle on the cycle.

Awareness of others' time-use patterns

Now we shift the focus from increasing awareness of how you spend/invest *your* time, to becoming more aware of how other people spend *their* time. This is an interesting question in general, but it is actually essential for good leadership. Where do the people around you spend their time? What are the categories you would expect each of your direct reports to be including as they are spreading their time throughout the day and week? Do you think most of your people are even aware of their own time-use patterns on the job?

Well, the best way to find out is to ask. What we have found is that most people are not really conscious of where they spend their time. Some professions, like lawyers and accountants, bill their time by the quarter of an hour, but most of us are not that disciplined. Many people fill out time sheets with project codes to indicate time-use but they rarely provide insight into specifically what they were doing and not just what they were working on. Time as a resource has value and being aware of how it is spent is critical for high-performance leaders.

In most organizations, people are paid for their time. It is one of the most expensive resources in most organizations and needs to be "accounted for." It should not be how many hours or days a person puts in on the job, but are they spending that time on the right processes and focused from the appropriate level and time frame? A rhetorical question but a very important one. There is a constant problem of spending time on what is urgent instead of what is important. Too often leaders spend their time solving their direct reports' problems and not spending enough time dealing with the longer-term and structural issues.

Another critical point is estimating how much calendar time and level of effort the role requires in order to accomplish its objectives. Knowing how time is spent can dramatically improve future efforts to resource projects. And, too often, leaders tend to underestimate how long things will take.

Effective delegation is the foundation for good leadership. Delegation depends on a leader's ability to accurately estimate how much calendar time and work hours will be required to complete the task. Delegation is based on balancing available human resources to specific tasks' time requirements.

One day Richard started our visit with the question, "How do I get my developers to be more efficient, so it takes less calendar time and therefore less budget?"

I asked, "Where do they spend their time now?"

He started telling me the makeup of the teams and the roles they played but did not really answer my question. When I repeated the question, he looked up and then down and finally said, "I have no idea."

"Well, it seems to me that it is hard to figure out how to be more efficient if you do not know where time is being used now," I replied.

Another long pause and then Richard said, "We can solve this by getting the level of effort better aligned with our evolving priorities."

Richard was gaining valuable understanding of time-use patterns.

Time-use habits

Why do you think it is so hard for most of us to notice where we spend our time? Part of it might be because we do not see the value of creating that kind of self-reflection. But much of it is that our sense of time is primarily in our subconscious, and we are simply not aware in our objective consciousness. Again, using the meta-perspective technique we can bring those subconscious patterns into conscious focus. There is more to it than installing triggers to step back and reflect on time-use patterns.

Most of our time-use patterns are habits. We do not usually think about our habits. We simply follow along and repeat the pattern. All habits are like this, but mental habits are not as obvious as behavioral habits. We tend to simply punch the button and let our programs replay. Habits are generally stable over time and can create ruts that make it hard to change our "normal" patterns when priorities change.

Changing our time-use patterns to be more productive and/or enjoyable requires that we examine our time-use habit patterns. Time-use patterns need to change when priorities change. Many times, even though someone knows their priorities have shifted or new work has been added, they have a hard time changing their "normal" time-use habits.

Time of attention vs. time on task

When we are trying to be more aware of our time-use patterns, it is easy to get confused between actually working on a task and where our minds are actually focusing our attention. Time of attention is how much time, and how often, does your mind shift to other concerns and priorities while we are supposed to be working on something else.

Attention is the focal point of consciousness. It determines what we are thinking about at every moment. At all times, moment-by-moment, our attention is focused on something. Consciousness always has some object of awareness. Even when people space out, if we query them afterwards, we would discover that indeed they were thinking about something, even though they weren't consciously aware of it at the time. They may not have reflected on their stream-of-consciousness, yet they were, nevertheless, attending to something, even if it was vague and free-flowing. We can be aware of many things at the same time, but we pay attention to only one thought at a time.

Awareness is like sunlight, illuminating everything in its field. Attention is like a flashlight, focusing in on a small part of the bigger picture. We can focus the beam of light on any topic, but it is always focused on something. We can focus our attention on any temporal domain. You can point it on the path ahead (your future), shine it on your moment-to-moment experiences as they are happening (your present), or focus it behind you, on memories (your past).

An advanced skill of TQ is to learn to notice the patterns and habits of what we pay attention to and how focused our attention is over time. With intentional effort, you can develop the ability to be self-reflective and observe your own patterns of attention, and then to change those patterns to make yourself more productive, more fulfilled, and more joyful. Time-of-attention skills, however, are not only about where and how you point your *flashlight of consciousness*. They also involve learning how to concentrate the power of your beam—how to *focus attention over time.*

I had a stark experience of this distinction in my own life. I had to be at a conference in Dallas where I was to be a speaker and I was not happy about wasting big chunks of time listening to lectures and aimlessly networking. I had a big project emerging and I really needed creative time to prepare for an important presentation the following week. After finding myself distracted during what was actually a very interesting talk, I realized that I had been focused on the new project. What was interesting is that I noticed that I had had a breakthrough in several of the key design issues I had been grappling with. Even though I had only spent a few minutes thinking about the issues, I had come up with some very creative ideas. That is when I started making an intentional effort to appreciate the difference between time of attention vs. time on task.

Leadership Practice: Comparing actual time-use to target time-use

Description:

This practice builds on the practice from the previous chapter where we discussed establishing a Time Chart of Accounts and Budget. Developing a time chart of accounts and budget is similar to establishing a financial budget. The next step is to compare actual time spent against the time budget. This practice provides a framework for reflecting on time-use in a particular time window and provides a basis for analysis and adaptation. The objective is not to strive to meet the target time-use but to gain insight into both potential areas for improvement and areas where expectations need to be adapted.

One does not need to keep a detailed time sheet to be able to do this practice. Rough estimation is sufficient, although increased awareness is a foundation TQ capacity. This is a practice to become more aware of your primary time-use patterns. Once you have established a set of time-use categories, and have reflected on your time-use patterns, you can move toward your ideal balance over time.

Process Steps:

Step 1 – Use the time chart of accounts and budget developed in the last chapter. This is often a challenging exercise because we are so often unclear on how we could best spread our time use. When asked where we spend our time, we should be able to answer. This is the purpose of a time chart of accounts.

Step 2 – Using the framework established, track your time-use in each category and major sub-category. Again, every week is unique with shifting priorities and business cycles, but our patterns can be observed if we have a framework to help categorize our time-use.

Step 3 – Begin to record your actual time-use. Take time every week to fill in actual time-use and analyze where you are over and under your targets.

Step 4 – This is an ongoing step to notice variances from target and either adapting how you spread your time or adapting expectations.

Practice Opportunity Triggers:

This is a practice that should be attended to on a regular, disciplined schedule. Ideally, a person would take three minutes at the end of each day and record the daily experience. Then weekly, a person should review the time-use pattern for the week and either adapt priorities or adapt the time budget.

When I was an executive and had direct reports, I required my people to keep track of their time-use patterns. There was always push back, but once a young rising star said, "I hate keeping track of how I use my time, but I am sure glad I do. It keeps me focused on the right things."

TQ Organizational Development Implications

Awareness of time-use is not only a matter of personal effectiveness, it also has significant potential for improving organizational design and performance. It is very important for you to know how you and your people spread your time in different roles, functions, and projects. It is also important to understand the level of complexity and the time frame of the problems that you and they are spending time on, based on the different organizational levels. (Fig. 2.)

Organizational Levels & Time Span of Role
Statified Systems Theory

Maximum Time-span of Role	Organizational Level or Strata	Industry Roles	Army Ranks
50 Years	Strata VII	Holding Company CEO	Five Star General
20 Years	Strata VI	Holding Company EVP	Three to Four Star General
10 Years	Strata V	Business Unit President	Division (Two Star MG)
5 Years	Strata IV	General Manager	Brigade (One Star BG General)
2 Years	Strata III	Unit Manager	Battalion (Lt. Col./Major)
1 Year	Strata II	First Line Manager	Company (Capt./Lt.)
3 Months	Strata I	Operator/ Staff	NCO and Private

(Fig. 2)

When each individual's time and attention is focused on the tasks appropriate to their mental capacity and time frame, what professionals call "working at the top of their license," it increases the probability that the organization can execute more effectively, and that each individual is achieving personal satisfaction. The means to those outcomes are effective delegating practices based on an awareness of time frames and time-use.

It is common for many organizations to take a casual and unstructured approach to the most fundamental practice in any managerial hierarchy, transmitting the intention of the governing body from the CEO down to the front line. When conscious practices based on an awareness of time frames and time-use are introduced, it can be startling to see the transformation in both organizational and individual performance. Much of what passes for organizational friction is a result of casual or poorly conceived delegation of time usage.

The point of the above table is that in order to have a "requisite organization," it should be designed to fit the complexity and time frames of the relevant work processes.

Reflections and Actions

Rate your current level of awareness of where you spend your time at work on a scale of 1 to 10 (10 high). Are you satisfied with that level of TQ? Make a commitment to work on increasing your awareness of time-use patterns.

- "Your time is your most valuable asset. How you spread that asset over multiple priority categories determines how effective you will be in your leadership role." How do you react to this statement? Do you consider your time your most valuable asset? What is your ROI on time investments?

- "In most organizations, people are paid for their time. It is one of the most expensive resources in most organizations and needs to be "accounted for." It should not be how many hours or days a person puts in on the job but are they spending that time on the right processes and are focused from the appropriate level and time frame? In what ways do you think time-use is not really the key issue. Just spending time does not really count, but you must spend time to get things done. Is it more important to be efficient with your time or effective with your time?

- What are some of your most routinized time-use habits? Notice how much of our time is spent according to unconscious, unintentional habits. Identify one time-use habit you want to change.

- Do you make a distinction between time spent and time invested? What makes the difference? Can you think of some accomplishment where you invested a great deal of time to reach that goal? Was it worth it? What are you investing in now?

CHAPTER THREE

Estimating Time-Requirements

How well do you "budget" your time? How well do your people budget their time? How clear are your team members on where you expect them to use their time? How do your people balance their priorities and spread their limited hours in a day, month, year? As discussed in the last chapters, most of us are unaware of where we spend/invest our time. And most of us are unaware of how we can estimate our expectations of the time a particular event will take.

TQ skills include improving our ability to accurately estimate the time a process or a cycle will take. Time-requirement estimations are a calculus combining calendar time, human resource time, and the pace of external dependencies. Adding human resources to a project may not speed the calendar time it will take to complete it. And shorter calendar time frames without an adequate level of effort from key resources will result in not meeting the estimated calendar time. All significant processes are embedded in a broader system, creating a wide range of interdependencies.

We all use time-estimating mental processes every day. Our mental calculations of when we expect some "thing" to occur are generally on a subconscious basis, but we can actually pull those time-estimating formulas into conscious focus and adapt them. In addition, we can add feedback loops into our mental processes to be more agile when circumstances change. Refining our "time-estimation algorithms" is intentionally installing feedback loops into the formulas we use to predict the timing of particular outcomes from events, processes and cycles.

Thesis: The thesis of this chapter is that it is important to be aware of how we individually—and our managers and direct reports—are at estimating what level of effort in human-resource time it will take in calendar time to complete a task, project, or work cycle, given its dependency on the timing of external variables. Based upon this awareness, we can improve the effectiveness of our mental calculus and algorithms by developing TQ for estimating calendar time and resource time allocations to be more effective. And good leaders help their team do the same.

A basic management capacity is to effectively forecast when key milestones and outcomes will happen. It is difficult to describe how important it is for leaders to be able to effectively estimate how much human-resource time and calendar time is necessary to accomplish results—on time, quality, and budget. As noted above, in most cases, the mental calculations a person goes through to predict the time-requirements estimation process is subconscious. Not surprisingly, these are often not well thought through. And when we extend this to complex organizations, the complexity is overwhelming and the implications for success of the organization are daunting.

The challenge to improving TQ for time-estimating capacity is to notice the accuracy and tendencies of our "time-use estimator". This means being intentional about what factors we are considering in the mental time-requirement-estimation process. By observing our own time-estimation patterns over time, and the way those that report to us (not to mention those we report to), we can dramatically improve not only our ability to better forecast outcome time frames, but also to improve our ability to delegate the appropriate level of resources for the task or project.

TQ Concepts to be covered in this chapter:

- Default assumptions for time-estimation
- Estimating the level of effort and calendar time

- Estimating time implications of "traction" and "friction"
- Observing the time-use estimation patterns in others
- Time-requirement estimation algorithms

Leadership Practice: Monitoring and refining our "Time-Estimation Algorithms"

• • • • •

Case Study

In my conversations with Richard and his need for more with less, we shifted the focus from an imperative to cut costs to looking at the set of competing priorities among different senior leaders. The investments most important to some departments were not valued at all by others.

Richard realized he needed to separate the different value propositions for different leaders. This opened a whole new way of thinking about his dilemma.

First, I asked him, "Do the senior executives you work with know how your people actually spend their time?"

When he stumbled, I followed with, "Do you really understand how your people actually spend their time? I know you have time reports, but all of your budgets are based on assumptions you have made about the level of effort required to meet priority objectives."

Then I shifted to what I wanted us to talk about, "How do you estimate time requirements against specific outcomes?"

Richard paused again so I knew he needed further understanding of mental formulas in time-requirement estimations.

TQ Concepts

Default assumptions for time-estimation

We all have a set of default assumptions about how we expect the future to unfold. The world is too complicated to track all the moving parts and therefore we need to base much of our decision-making on assumptions. Most of the time our assumptions have been built into simple default settings with no further analysis for potential change or accuracy in a particular situation. The challenge of increasing TQ in this area is to bring those subconscious default assumptions into focus and more intentionally analyze the sequence of factors that lead to an outcome.

Default settings are not a bad thing. In fact, creating default settings for our internal programs is essential for cognitive efficiency. The entire way we construct our understanding of the future is based on assumptions about the past and the present and we then project those assumptions into the future. The challenge is to refresh those assumptions as things change and to assess them considering the specific challenge under analysis. It is common, and good practice, to develop mental shortcuts and routines, but too often those useful shortcuts turn into habits and become unexamined, even unexaminable, assumptions.

In this area of TQ development, examination of our internal default time-estimation settings is key to improving our ability to accurately forecast outcomes. We need to identify our default assumptions and examine the validity and usefulness of those assumptions. In addition, it can be extremely valuable to get other people to articulate their mental calculus for how they determined anticipated time frames.

Estimating the level of effort and calendar time requirements

Time estimation mental calculation has two interdependent variables. First is calendar time, or how long something will take in terms of hours, days, weeks, etc. The second is how much time as a resource is expected to be applied to getting something done. These variables are interdependent, but not always predictable. The relationship between the calendar-time expectation and the level of effort in terms of time from key resources is a tricky dynamic.

One of the major actors affecting the accuracy of one's time estimation is other competing demands for key resource time. Time estimates made on a wide variety of external variables are always made with assumptions built into the relevant context. When those factors outside of one's control change, or our assumptions about those factors are inaccurate, it will affect the time of completion.

Likewise, identifying and securing the human-time resources required to complete a task or process is directly affected by the quality and availability of key resources. Most business processes involve multiple people in complementary roles, with each performing a necessary step in the process. Inadequate, or unskilled, human resource time is a common cause of a project running over the original deadline.

After about a week of trying to reframe his dilemma, our CIO friend, Richard, found himself faced with another dilemma. How good were his timing and budget estimates? When the annual budget had been agreed upon almost ten months earlier, he had submitted his plan, assuming resources would be available as they always had been. That was before the acquisition. Now he had to go back and really scrub what gains could be made by refining the project plans.

He was trying to justify his budget requirements by linking them with results that different senior leaders wanted, and I hit him with a question that stumped him again, "How clear are you about the mental calculus that went into estimating the time requirements for these projects?"

"What do you mean?" he asked.

"We all do a mental calculation to determine how long a project will take to complete. And we have individual tendencies. Some people almost always underestimate time requirements, while others almost always *over*estimate."

"I have certainly seen that," Richard added, "so, what do we do?"

I shifted tone and said, "Well, it is fairly simple, just not easy. We need to walk through your work plans and budgets and see if there is any room to tighten or shift priorities."

Estimating time implications of "traction" and "friction"

There are two different forces that affect how long something will take. "Traction" is a description of when a project or process is connecting with key stakeholders and getting decisions made. How long it takes to get the sequence of decisions made is hard to estimate, but it is often the deterring factor on time-estimation accuracy.

"Friction" is a metaphor for nonalignment, resulting in personal and/or organizational conflict. A great deal of energy and time are wasted on addressing and resolving the drag of social friction. Friction is also often one of the primary factors in why things take longer than you think they will. It is an indicator or signal that things are not aligned. If everything is running smoothly, milestones are reached, according to plan.

On one occasion, I was called in for what was described as "a turnaround" situation. Everybody was pointing the finger at everybody else, and I was asked by an old colleague to try to help. As usual, things seemed fairly straightforward on the first pass, but I began to see the problem. What had started off as power dynamics between departments had turned into a set of *personal* problems. People were really upset with each other. Going way beyond fighting for their department's priorities and concerns, the personal baggage was causing a lot of friction.

Everything took longer than it should because people were not cooperating. In fact, they were intentionally slowing things down. By reassigning one key individual and a few timely conversations with others, soon the friction had gone away. I sometimes call it "emotional grease".

Observing the time-use estimation patterns in others

As we have done with previous TQ skills, after developing awareness of our own temporal patterns, we are able to shift perspectives to focus on the temporal patterns of others. In this case, our focus is on how effective one is in accurately estimating when something will happen and how much time resource needs to be spent on that outcome.

At one level this is very common and easy to forecast. There are some people you know who are always optimistic. They dramatically underestimate the calendar time and level of effort required to get something done. On the other hand, there are some people who always tend to overestimate how long something will take or the level of effort they need. Sometimes, this is just sandbagging, but more often people that overestimate build in significant time for possible disruption of the process. Neither is good or bad, but it is valuable to notice how well others estimate time requirements and factor their tendencies into our mental algorithms.

One simple way to monitor other people's time-use estimating skills is to ask them. This is another case where simple TQ related questions can clear up a lot of misunderstandings and frictions. Most people are unaccustomed to having their time-related thinking questioned at all. They may rarely consider their own thinking and assumptions about time.

It has been our universal experience that when asked, everyone is intrigued by the question and the most common response is something along the line of, "Huh, I hadn't thought about it that way before." The resulting conversation is typically regarded as interesting, not challenging at all. Also ask them to explain the logic or formula they use to set their time-requirement expectations. Experience is often used as a means to calculate our mental algorithms with a lot of data. We can usually learn from seasoned people who have seen similar processes many times and know what variables to notice.

Time-requirement estimation algorithms

The TQ difference between typical subconscious, mental calculus is the intentional inclusion of feedback loops, or mental algorithms. Algorithms add a dynamic to "expectation formulas". Mental algorithms work better for forecasting the future than mental calculus. Effective algorithms are based upon clarity of critical-path benchmarks and interdependencies to give a leader the ability to constantly adapt to unfolding circumstances, including adjusting time-of-completion estimations.

A critical path is the sequence of key events that set the time constraints for a project to reach a successful outcome. The critical path usually consists of key deliverables, key decision points, and key external factors. Estimating the final completion time can be calibrated by observing the timing of key benchmarks or milestones.

When a project plan is considered on a linear timeline, key benchmarks are needed to mark the different phases and stages required to complete the overall project. Like running a mile relay, the final time is determined by the time required for each leg of the relay. When there is a delay in one of the cycles that is nested in the overall time frame, it often has implications for the timing of the final outcome. With greater insight into the critical path benchmarks, a leader is better able to adjust both level of effort committed and timing estimations.

Of course, this assumes that a leader can get information on a timely basis about key benchmarks and potential delays. This is where effective delegation can establish the variables and indicators that help gain insight into the timing of a longer-term project. By monitoring the timing of milestones, a leader is able to adjust as necessary to adapt to the changing circumstances. Everyone on our project should be clear about the critical path benchmarks and the timing required to meet each one in order to meet the overall timing of the project.

This situation presented itself several times after the COVID-19 pandemic. When the global network of just-in-time inventory systems all shut down at once, the timing expectations for supply chains crashed. Without good information about delivery times of raw inputs and components, time estimates will go haywire. This called for a broad re-evaluation of inventory and delivery time requirements and expectation.

Leadership Practice: Monitoring and refining "Time-Estimation Algorithms"

Description:

This is a challenging practice. All of our mental algorithms run in our subconscious mind and are difficult to unravel. Focusing in on time-estimation algorithms is an advanced TQ capacity, but the benefits can be significant.

As the name indicates, there are two major parts to this practice, monitoring and refining.

Process Steps:

Step 1 – Identify a major project you are working on and put it into a strategic frame that notes context and outcome expectations. Bring your conscious focus to note your assumptions built into the timing of the expected outcome.

Step 2 – Starting with the outcome and its expected time frame, back up through the entire process required and the expected timing, and identify the key benchmarks. Bring to your attention the assumptions built into your "expectation model." Make a note on each key target date.

Step 3 – Monitor the accuracy of your original time-requirement estimates and notice when they were too long or too short. Analyze your estimation factors and expectation model and sort for potential reasons for over- or under-estimation.

Step 4 – Note the factor and build in feedback loops to adapt "default estimation assumptions."

Practice Opportunity Triggers:

When delegating a task to someone in your organization, ask them to tell you how they calculate the time required. Then think through how you came up with your timing forecast.

Each time we set a schedule for a process, use that as a chance to reflect on how you determined your estimated time.

TQ Organizational Development Implications

Time frame capacity

How far into the "future" someone can create and maintain a mental model of a system dynamic depends upon the complexity of the system and the cognitive capacity of the individual. The "future" is a psychological domain of consciousness and does not really "exist" except in a person's extrapolation of probabilities. Research has demonstrated that different people have different levels of cognitive capacity that reflect their different levels of potential time horizons.

Based upon well-documented research, Elliot Jaques proposed a model that reflects growth in cognitive capacity and potential time frames. (Fig. 3.) This stratified model is shown on the next page.

Jaques has gotten significant criticism for his observation that cognitive capacity is demonstrated by time frame capacity. Not everyone's mental computer has the same RAM storage. Understanding your own cognitive capacity to project a coherent version of the unfolding future is the first step toward TQ.

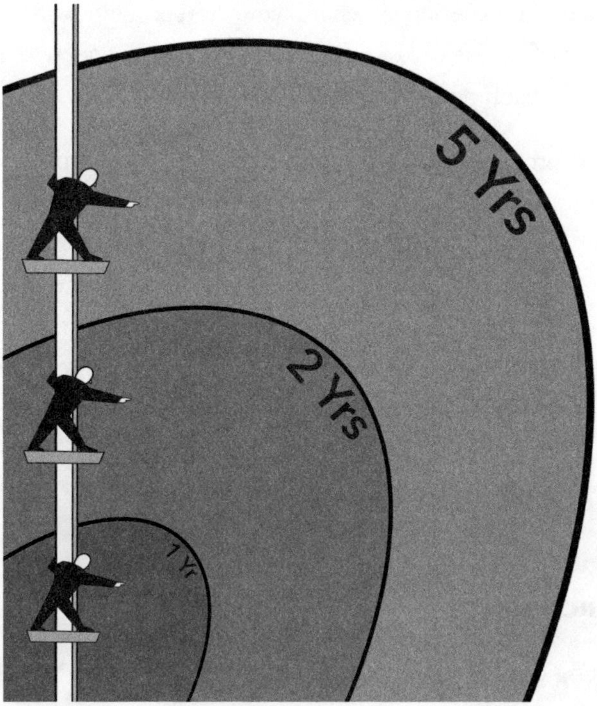

(Fig. 3)

Reflections and Actions

- How do you assess your time-estimation skills on a scale of 1-10? Can you see the value in developing this dimension of TQ? What is a specific practice you can use to improve in this area?
- "In this area of TQ development, examination of our internal default time estimation settings is key to improving our ability to accurately forecast."

  How aware are you of your default settings? What can you do to become more aware of your internal default time estimation settings?
- "The relationship between the calendar-time expectation and the level of effort in terms of time from key resources is a tricky dynamic."

Have you had an experience where your target date was missed because of lack of access to key resources? Next time, review the time requirements for each person involved.

- Two different forces can affect timing: traction and friction. Do these two metaphors make sense to you? Can you recall a time when you were working on a team that had great traction and made great progress? On the other hand, have you been in a situation where there was a great deal of friction? How did that friction impact the quality and timing of the deliverable? What can you do to factor in traction and friction into your timing estimates?

• • • • •

Case Study Conclusion

So how did increasing his TQ help Richard Smoot deal with his "unsolvable problem"? Of course it is always a challenge to tell a long story in a short amount of time, but in this case, the TQ approach really helped him resolve his dilemma.

If you will recall, he had conflicting demands from different senior leaders and was vexed by the "impossible expectations". Based upon multiple conversations, he was able to get the different leaders to understand their priorities in the context of the broader corporate challenges. By helping the senior executive team to understand the conflicting expectations, he was able to get them to push for their interests.

In a big senior management meeting, he presented the conflicting expectations not as a "problem" but as a set of trade-offs that had to be made. By getting each of them to present their needs, time frames and resource requirements, he was able to get them to sort it out instead of leaving him in the breach.

This case study is a common situation. Different executives have different priorities with different resource requirements and time frames. Instead of being wrong no matter what Richard had decided was the "best" mix, he was able to get them to make those tough decisions.

And in the end, he got more money, not less. And maybe more importantly to him, everyone gave him very positive feedback on his leadership instead of blaming him for making the tough priority trade offs.

PART TWO

Time as a Frame of Reference

INTRODUCTION

In this part, we will shift from thinking about time as a resource as discussed in the last part to focusing on how we use time as a frame of reference. Our frame of reference determines our context and point of view. A temporal dimension is always embedded in the "relevant" frame of reference we use to make sense of the dynamics in our lives. Time frame is the most fundamental aspect of TQ. Time as a resource, time-usage and timing are all aspects of time frame and effective practices in these areas are all informed by an awareness of personal time frame and the time frames of others.

This part deals with how we use time as a fundamental part of how we frame our perspective on our moment-to-moment experience. This uses **chronological time** named after the Greek god Chronos to calibrate our frame of reference. This refers to linear time—a timeline constructed of temporal units in a sequence. Time frame creates a mental container for our projection into the future and memory of the past.

A timeline is a theoretically infinite extension of temporal units into the past or the future. Time frames are chunks of time, or the projected temporal frame of reference we use to make sense of our past experience and to forecast the uncertain future. On the infinite timeline of sequential temporal units, we bracket a segment of temporal units into a "time frame". This segment of time provides the context, or frame of reference, to understand and respond to the events, issues, and problems that make up our jobs.

In the general Introduction to this book, we introduced the concept of emotional intelligence (EQ). Teaching and training EQ had been a part of my work for a long time. As I began working with temporal skills and concepts, I realized that the developmental levels and challenges are very similar to those of emotional skills.

Here we will be leveraging Daniel Goleman's levels of emotional intelligence to create a framework that can help people grasp the key concepts and improve their temporal intelligence and skills around time as a frame of reference. The four developmental levels of emotional intelligence (EQ) are paralleled in the four developmental dimensions of temporal intelligence (TQ) associated with time-frame awareness and agility.

Below is a representation of the relationships between EQ/TQ levels of development:

Summary Table for the EQ/TQ Levels of Development

Dimensions of EQ & TQ Development	EQ Description	TQ Description
Self-Awareness	Aware of own *emotions*	Aware of own *time frames* of analysis
Ability to adapt	Adapt own *emotional* reactions/responses	Adapt own *time frames* of analysis
Awareness of Others	Aware of others' *emotions*	Aware of others' time frames
Ability to Influence Others	Help others adapt their *emotional* reactions/ responses	Help get in sync with others to create a common *time frame* of analysis

Understanding and developing TQ with time frames is a very powerful leadership capacity. Time frames are the subjective temporal context within which we analyze and make decisions. Time is always a dimension of the frame of reference we are using when working on a particular problem or opportunity. The power to control **context** is much more important than to control **content**.

The power to control **context** *is much more important than to control* **content.**

Time-frame management in an organization allows individuals to work with focus and intensity on their goals. You can't ask someone to focus on an intention beyond their time frame, but you can organize a progression of time frames to achieve your longer-range outcome.

Stephen Bungay in *The Art of Action* describes this as "Direct Optimism". Providing strategy and policy guidance to enable everyone to work collaboratively inside their appropriate time frame. Jaque's insight is that different individuals have different time-frame capacities. The military explicitly briefs and back briefs officers, noncommissioned officers and soldiers in relationship to their time frames of action. Back briefings are a mechanism to convey time frame information upward. Strategy and policy communicate time frame information downward and laterally. This is an example of practices developed to explicitly convey time-related information in operating environments where the margins for error are life or death.

In our work some of the most common problems we encounter are unconscious time frame confusion or conflicts. These proliferate everywhere and are a source of unnecessary frustration throughout hierarchies. The problem is sometimes attributable to poorly designed hierarchies which incorporate the design flaws that Jaques spent his career cataloging: overlapping time frames, overly distant time frames, or overly close time frames. However, the more fundamental source of difficulty is a lack of TQ and an awareness of the significance of time frames in general.

Once someone is sensitized to the significance of time frames and the utility of time frames as an objective function, it's relatively easy to diagnose the presence of a time frame-related issue and individually adjust their work to accommodate any hierarchy design issues. A time frame-aware executive with high TQ can still function effectively in a time frame-confused hierarchy.

• • • • •

Case Study

As we did in the first part of this book, in this second part we begin with a case study from our experience that highlights the issues addressed in each chapter. These scenarios will then be used to help us analyze and reflect on TQ in practice.

This case study is based on some executive coaching and assessment work done several years ago. Craig Mack was the CEO of a mechanical engineering firm operating in the Seattle area. Craig's father had started the company thirty years earlier "with a pickup truck and two friends from his high school football team." Over time the company grew into one of the larger firms on the West Coast. Seven years ago, Craig and two of the other executives bought the firm from Craig's father with some cash and a large note to his father. I had been introduced to Craig by a friend, his Vice President of Human Resources. As we sat in Craig's office, this is how he began to talk about his situation:

"When my father sold us the company, we were in good shape and ambitious. For the first five years we had good growth, but it turned out that it was less about us and more about the market. Now things have slowed down and we're getting into trouble. The quality of our work has stagnated, especially compared with our best competitors. Our costs are getting too high, and our sales team isn't technically sophisticated. Clients and general contractors are coming to expect broader expertise in design and engineering. We're good at getting in and getting work done quickly, but we don't have the overall design expertise that our larger competitors bring to the table.

"If we don't change, we're going to be out of business. My father lent us the money to buy the business and my partners are starting to talk about renegotiating our loan to give us some breathing room. I'm not going to let that happen. The problem is that it seems as if we must change everything, and we can't agree on when to start what."

I could tell Craig was caught in his own circular logic. He was trying to sort out one problem but then realized that each problem was interdependent and overlapping in time with other problems.

"Where and when do *you* think you should start?" I asked.

After a moment of silence, I suggested we explore the four time frame levels of TQ development.

We will continue to use the conversation with Craig throughout Part Two. This case example provides a good vantage point into the key concepts covered in each chapter. The chapters and scenarios from Craig's story help answer the following key questions:

- How aware was he of the different time frames he was using to understand different problems or opportunities?
- How effectively could he shift from one time frame to another to give him a different perspective?
- How well did he understand the time frame being used by others when working together?
- How could he get other key leaders in sync around nested time frames and overlapping dependencies?

Overview

There are four chapters in this section reflecting the four levels of TQ development. Each chapter provides a conceptual introduction, key TQ concepts, leadership practices for applying TQ time frame skills, and their organizational implications.

Chapter Four - *Self-Awareness of Time Frames*. We can all benefit from developing greater conscious awareness of the different time frames we use when dealing with different parts of our jobs. The key challenge for TQ Dimension One is to intentionally become more aware of the time element of the frame of reference applied in everyday situations. Shifting perspective, or as we are calling it a "meta-reframe," gives us the ability to observe our own moment-to-moment thinking and adjust to a different frame of reference.

Chapter Five - ***Agility of Time Frames***. This chapter is focused on the second dimension of TQ—developing greater agility and adaptability of your own time frames. Awareness is important, but agility is valuable.

The ability to intentionally shift one's time frame of analysis to different perspectives has numerous benefits.

Chapter Six – ***Awareness of Others' Time Frames***. The third level of TQ is about noticing another person's time frame of analysis. Noticing something, anything, requires us to focus our attention and sort for the "thing" we are looking for. We only notice things, or patterns, when we pay attention and focus. Increasing awareness of others' time frames requires the interplay between reflective observations and inquiring questions.

Chapter Seven – ***Ability to Sync with Others' Time Frames.*** This chapter focuses on one's ability to get in sync with the time frame being used by others when dealing with a specific problem or opportunity. As mentioned above, getting in sync, or getting in alignment, includes both having awareness of our own time frames and being able to lead other people to share a common time horizon for specific expectations.

New concepts and practices are introduced in this section that are interesting, but the real value to you as a leader will be determined by how you work with these concepts in daily practice to develop your leadership capacity. It can be a challenge to intentionally take a meta-perspective and reflect on the context created by the time frame used during real-time analysis at a decision-making meeting.

CHAPTER FOUR

Self-Awareness of Time Frames

Like emotional intelligence, temporal intelligence begins with self-awareness. The foundation for developing greater TQ begins with our ability to be aware of our own time frames.

Dimensions of Temporal Intelligence (TQ)

Level of TQ Development	TQ Description	TQ Leadership Practice
Self-Awareness	Aware of one's own *time frame(s)*	Meta-Perspective

Temporal intelligence includes becoming aware of the time frames we, and others, use while dealing with any specific situation, problem, or opportunity. This first dimension may seem like a simple exercise, but most people find it difficult to shift their awareness from the contents of a situation to the frame of reference. Time frame is always an element in a frame of reference. Level One of TQ development is increasing awareness of the time frame of reference being used at any moment to deal with the issue or problem at hand.

We can all benefit from developing greater conscious awareness of the different time frames we use when dealing with different parts of our jobs. The key challenge for Level One TQ development is to intentionally become more aware of the time element of the frame of reference applied in everyday situations, in fact at any moment. Shifting perspective, or as we are calling it "meta-reframe," gives us the ability to observe our own moment-to-moment thinking and adjust to a different frame of reference.

Thesis: The thesis for this chapter is that awareness of time frames has numerous practical implications, and this awareness can be developed by intentionally working on improving this capacity. It is one's ability to shift perspective and step outside of the current focus on the content and observe the time frame that is being applied to "understand" the situation. Again, at first this may seem simple but the value that comes from being aware of time frames has significant impact on one's effectiveness as a leader. Clarity of our own time frames is an essential element in all of core leadership skills, such as planning, communicating, delegating, and evaluating.

All of these core leadership competencies can be improved with awareness of the time frame one is using to analyze a particular problem or situation. The difficulty lies in the fact that they are typically subconscious. Our consciousness acts like the tip of an iceberg above the ocean surface, while our subconscious is like the bottom of the iceberg. Our subconscious "worldview" is built on our experiences, beliefs, and values. These are not typically in our conscious awareness but create the frame of reference we use when we are trying to make sense of a particular subject or situation. Bringing our temporal context into conscious focus is dipping into the subconscious and bringing one's *time frame of analysis* into conscious awareness.

TQ Concepts:

- Time frame of analysis
- Time frame of intention
- Time frame complexity and capacity
- Content/Context/Point of View

Leadership Practice: Meta-Reframe is a technique that provides the ability to step back and observe our own thought patterns, which makes it easier to be aware of our subconscious time frames.

• • • • •

Case Study

Craig explained, *"Look, what's important to me is that we keep this company going and keep our commitment to my dad. He trusted us with this and he's counting on us paying him back. That's a ten-year commitment and we have to pay it off over that time frame. I keep saying we should attack one thing at a time and start on our quality and sophistication. If we're not close to the best, why would anybody want to hire us?*

"Once we have competitive services, we can focus on sales, but for now, the salespeople can be involved with the quality initiative. Our new engineers, and our fabricating and installation people can work with sales to build skills, selling tools, and communications links back to the engineers and fabricators. I think we should leave the cost-cutting to the last. It's all about timing. Without the sales force to sell it, the cost-cutting wouldn't create value for anyone."

Then I asked Craig, "You mentioned 'time frame' a minute ago. What time frames are you using here?"

"That's a good question," Craig said. "I guess I hadn't thought much about it."

"I know the question can be confusing. But when we do think about it, we realize we deal with different issues with different time frames. The starting point is to be clear about the time frame of the primary problem you are working with. What is the time frame of your primary objective?

TQ Concepts

Time Frame of Analysis

Time frame of Analysis is an important concept in TQ. Every time we "analyze" something, we must have a temporal frame of reference, or a time frame. Understanding any content requires that we construct a context within which we make sense of the information available. Focusing on a particular situation will always include a time frame embedded in the context. The time frame used to understand a problem is the ***time frame of analysis***.

The time frame used to understand a problem is the time frame of analysis.

This will be a key term used throughout this part of this book to mean the temporal frame of reference used to analyze and understand a situation, problem, or opportunity.

Time frame of analysis is the temporal context projected into the future and cast back to the relevant past. The subjectively constructed probable future is a function of the subjectively filtered past causes and trends. The longer one's time frame of analysis, the more complexity and ambiguity are brought into the way the future is extrapolated.

Time frame of analysis is an empty container of a certain temporal proportion. How that temporal context is filled with future probabilities and intentions is a different thing. Time frame is the temporal space within which we project our expectations of what will emerge as the future.

Time frame of Intention

Time frame of Intention is a second foundational concept for developing TQ. Time frame of intention recognizes that intentions for future outcomes always have a time frame. When a leader delegates an assignment, there is always an expected time in the future for the outcome to be accomplished. Unfortunately, the specific time frame expected may, or may not, be clearly communicated by the leader or understood by the subordinate.

Time frame of intention refers to the amount of linear time you expect for a task, or outcome, to be completed. The focus of a delegated task is often *what* is expected, instead of also clarifying *when* it should be done. Expectations always imply time frames.

Elliott Jaques uses the term "time horizon" to mean the length of time required to meet the expectation of a specific task. He acknowledges how we use different time frames for problems of different complexities.

Setting expectations in an organization is a basic requirement for leaders. All roles and responsibilities have some sort of time frame. However, in general, we are unaware of the time frames we are using; too often, so are others.

A classic story about the Alligators and Swamp captures some of our own experiences:

> *A person is sent to drain a swampy area. However, he discovered alligators in the swamp that he had to deal with before he could refocus on his original objective. As he handled one alligator, another would jump up. After a time, he was so busy fighting alligators that he had forgotten about draining the swamp. This is an example of a person who gets so busy fighting alligators, or dealing with short-term challenges, that he forgets about the original objective, which was to drain the swamp.*

Be aware of the tendency to focus on urgent short-term problems and ignore longer-term objectives.

Be aware of the tendency to focus on urgent short-term problems and ignore longer-term objectives.

Content/Context/Point of View

This concept is captured in a triplet. Triplets are a set of three perspectives that occur in a causal sequence. One leads to another. Often our operating triplets are subconscious, and we develop mental habits that reflect the 3-part relationship. The triplet below and those discussed in other chapters provide logical steppingstones to greater TQ.

There is an important distinction between content, context, and perspective. *Content* is the object under analysis. Content is what we "know". Context, on the other hand, is the subjective frame of reference we use to "understand" and interpret what we know. Both of these elements of rational analysis are a function of the point of view, or perspective, of the observer.

Both context and point of view are subconscious when we are actively focused on the object, or factors, being analyzed or even observed. The context provides a background upon which the object can stand out in the foreground. Rational thought processes require a context to make sense of the things it notices. Bringing these subconscious dynamics into conscious focus opens up a whole new way of "understanding" and options around "perspective".

Getting back to our case study, after Craig and I discussed this TQ concept, he thought a moment and said, "Thinking about which time frame should set the primary frame of analysis, I guess I'm starting with the commitment to my dad."

He thought a moment and then added, "That's at least another five years."

So, I suggested, "Lock in that five-year time frame as the frame of reference for other issues and options. It is important for us to be aware of the time frame we are using to understand the issue, or situation, or problem. That awareness provides a frame of reference for more effective analysis and decision-making."

Key Practice: Meta-Reframe, Taking a Meta-Perspective

Description:

A meta-reframe is shifting our conscious attention to be observant of our own thinking. To step back and observe one's previous perspective from another level is meta reframing. We are often focused on the content of our thinking and are not aware of *how* we are framing that content.

"Meta" means a different level; so, a meta-perspective means shifting to a perspective that transcends and includes the previous one. This point of view helps bring the previous context into focus as the content within a higher order perspective. This ability gives us the capacity to consciously notice the time span used within our frame of reference.

Process Steps:

Step 1 – You need to acknowledge the importance of bringing awareness of your time frame into consciousness. You need to recognize that you are constantly using time frames to understand the problem at hand. It is valuable to have the capacity to step outside of our current focus on the content of the problem and notice the time frame being used to frame and analyze the problem.

Step 2 – Set intentions and triggers to pay attention to your operating time frames, or your time frame of analysis. Shifting to a meta-perspective will give you the ability to step back and notice the framework being applied in your specific situation.

Step 3 – Use the mental space created by taking a meta-perspective to notice the time dimension of the frame of reference being used to make sense of a situation is a valuable capacity.

The level of proficiency one has in the ability to intentionally take a "meta-perspective" directly affects a leader's capacity.

Practice Opportunity Triggers:

A trigger is something that causes a person to shift to a particular frame of reference. It is common to say someone said something that "triggered us". Many of these triggers are subconscious. On the other hand, we can "set triggers" to shift our focus to notice particular factors. In this case, we are identifying a trigger that helps you shift to a meta-perspective and notice the time frame of analysis being used to understand a problem. We will be suggesting potential triggers with each of the key practices discussed in this book.

A basic trigger to help you shift to a meta-perspective and notice your time frame is when you find yourself in a "problem-solving mode". This is when you realize you have a problem that has multiple dimensions, and you need to think "outside the box". This sense of challenge can be a trigger you to step back and notice your time frame of analysis.

Meta reframing is a technique to help us improve our ability to be aware of our subjective context, which always includes a time frame. Recognizing one's own time frame of analysis is a key skill for effective leadership. Taking a meta-perspective is simple, but not always easy. One person I worked with for several years began to call the insights he gained from being able to consciously shift levels of perspective and create a new frame of reference "aha moments".

This is how he described it:

"I call them 'Aha Moments'. I get caught up in discussions or arguments about a situation and lose my sense of perspective. Since I learned to trigger myself to take a meta-perspective, I can suddenly shift to a whole new frame of reference. I find that great creativity emerges when problems are viewed from a higher-, or longer-term perspective.

"During a heated conversation with a colleague, I could not remember to shift perspective and think about my time frame of intention. I only thought about it later. I realized that I had gotten frustrated during the conversation when our disagreement stemmed from using different frames of reference for the situation under consideration. That became my trigger to take a meta-perspective and reframe the situation under analysis. Almost every time I use this practice, I still get that "aha" when I remember to step back and take a different point of view."

TQ Organizational Development Implications

Adaptability and agility to move between time frames

This is a simple, but powerful concept that Jaques applies to organizational design and development. Time measures for the capacity of an individual to manage complexity in their work and human hierarchies organized around the time and complexity of the work that has to be done. In a workforce, frontline workers begin with the capacity to manage tasks that can be completed in a day or two; CEOs of large managerial hierarchies have to manage tasks that might take a decade or longer to realize. (Fig. 4.)

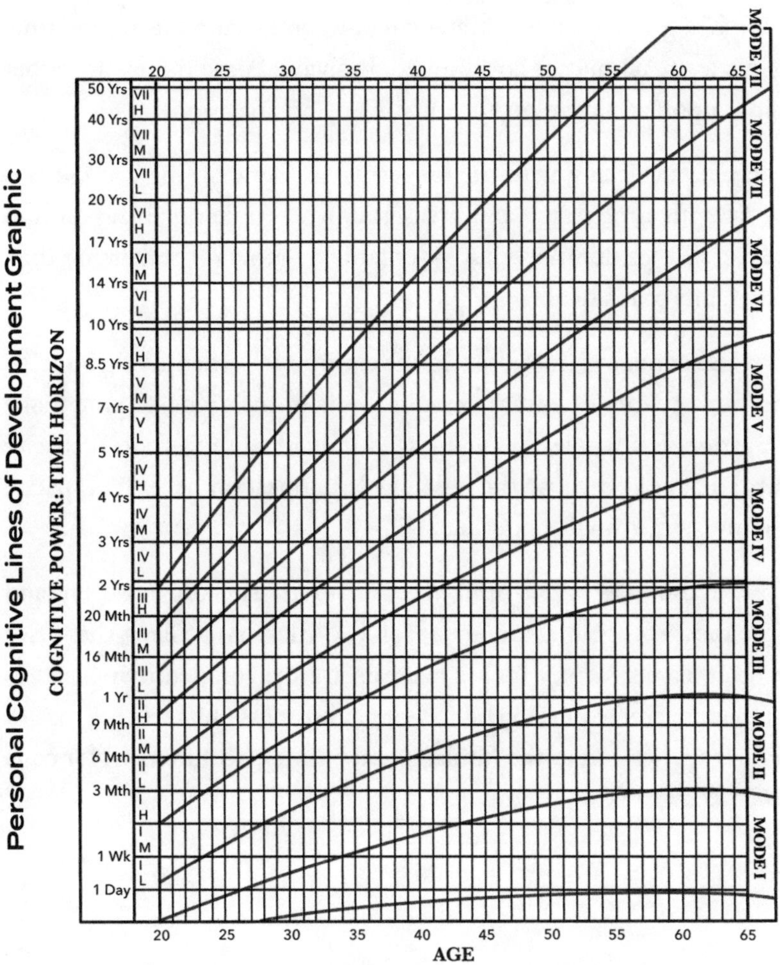

(Fig. 4)

The layers in a well-designed hierarchy, or "Requisite" in Jaques's terminology, sort themselves out around the time frames ranging from a day and decades. The previous chart illustrates Jaques's research on this organizing principle. You can take a few moments to think about your own longest assigned task that you're expected to pursue independently. Now you are in possession of a potentially powerful analytic tool based on four decades of research that can help you assess your work in relation to the people working around you.

In the case study that began this section, Craig Mack intuitively recognized this issue and consciously broke down the problem of re-engineering his firm into time frames that matched the capacity of his team. The key learning for Craig was that any strategy had to be executed using the time frames of his team, no matter how limited. Individual time frames are either a constraint or potentially an opportunity in any organization.

As the CEO working with the longest time frame on his team, it was his job to take his time frame and break it down for the organization into accessible, i.e. shorter duration chunks, organizing around the existing time frame constraints of his team.

The inverse of managing to time-frame constraints is identifying individuals whose time frames exceed the requirements of their jobs, or potentially the organization as a whole. Under-utilized individual time-frame capacity is organizational gold because the individuals in question have the capacity to drive innovation, growth, and value.

It is one of the most under-utilized resources in many organizations and perhaps the most compelling argument for TQ. Rich has a client, the CEO of a large healthcare system, who has made an art form of identifying leaders who have the capacity to develop billion-dollar business units. He is a master of TQ and its application to the fundamental business requirement for constant change and growth.

Reflections and Actions

- Reflect on the key concepts discussed above and think about a "takeaway" that you can use immediately. How can you put this new insight into practice?
- One of the essential points in this chapter is that time frames are normally subconscious. Do you agree with this assertion? What can you do to bring awareness of your time frame into conscious focus?
- Do you think you can improve your self-awareness of time frames by making intentional efforts? Reflect on a situation or problem you have experienced in the past where you had a short-term time frame when it would have been better to have a longer-term time frame.
- What kind of techniques and practices, like taking a meta-perspective, can help us step out of our normal frame of reference and focus on the time frame of analysis and intention?
- Since awareness of one's own time frame is an important leadership skill for planning, communication, delegation, evaluation, and other core capabilities, how can you insert this new insight into your regular leadership practices? A good start is beginning each meeting by writing WTF (What Time Frame) on the top of a piece of paper and noting the time frame you're using for the issue at hand and the time frames being used by others.
- What questions can you use to comfortably ask others about their time frames?

[illegible]

CHAPTER FIVE

Agility of Time Frames

The second dimension of TQ development is the ability to intentionally adapt, or shift, to different time frames. As discussed in Chapter Four, bringing our temporal context into conscious focus allows us to identify our own time frame of analysis. This capacity then provides the ability to intentionally modify our time-frame of analysis. It requires mental agility to shift from shorter to longer-term time horizons, or longer-term to shorter-term, when analyzing a problem.

Dimensions of Temporal Intelligence (TQ)

Level of TQ Development	TQ Description	TQ Leadership Practice
Self-Awareness	Aware of one's own *time frame(s)*	Meta-Perspective
Ability to adapt own	Adapt own *time frame(s)* of analysis	Tri-frame Reframe

Thesis: The ability to intentionally shift one's time frame of analysis to different perspectives has numerous benefits. As pointed out in the previous chapter, the first level of development for temporal intelligence is becoming aware of our own use of time frames when we're dealing with different situations or problems. This chapter is focused on the second dimension of TQ—developing greater agility and adaptability of your own time frames. Awareness is important, but agility is valuable.

Adaptability of time frames is the ability to shift the time frame of analysis to alternative temporal windows. Many great leaders seem to demonstrate this skill naturally. In fact, we all do it every day. However, we often get caught up in one perspective and have a hard time shifting to a longer or shorter time frame of reference.

The ability to shift one's own time frames can also be developed through intentional focus and practice. Building on the first level of awareness, adaptability of time frame provides great value on your effectiveness as a leader when planning, communicating, delegating, and evaluating. All of these leadership competencies can also be improved with increased agility of time frames.

TQ Concepts

- Generic strategic framework
- Nested Frames of Reference
- Set/Subset/Superset
- Time frame Reframe

Leadership Practice: Tri-frame nested reframe

• • • • •

Case Study

After Craig and I discussed practical examples of the value of him being more aware of his own time frame of analysis, I shifted the conversation, "What about some of those other pressures and problems you mentioned earlier? Have you thought about the different time frames you use to understand those different challenges?"

"Yes. I would think that the quality improvement effort will take a year, the sales improvements something like eighteen to twenty-four months, and the cost improvements about three years."

"Great," I replied. "So each of them is operating within different but interconnected time lines, and each of them needs to be integrated within the overall five-year window. So, in fact, you operate in multiple nested time frames. The challenge is to keep track of them and their interdependence by developing the TQ skill of intentionally shifting from one time frame to another and keeping track of the nested cycles."

Craig was nodding his head up and down, but his face seemed confused. "What do you mean?"

I asked, "How agile are you when it comes to shifting between time frames when focusing on interlocking different challenges or opportunities?"

After another pause, he confessed, "Sometimes I think I'm pretty good at shifting time frames, when necessary, but I do it more subconsciously."

Using 'air quotes' he followed up, saying. "I'm not really sure how to 'intentionally shift from one time frame to another' to get multiple perspectives."

"Let's explore that," I said. And we did.

TQ Concepts

Generic Strategic Framework

What is a "strategy"? What is the difference between a strategy and a tactic? Is a strategy the same thing as a plan? Well, it can be confusing, but I have spent years answering these questions. In a nutshell, a strategy is a framework for defining and solving a problem or pursuing an opportunity.

A complete strategy requires a framework made up of multiple essential components. When they are not well-articulated, it is hard to put the pieces together. The key point for our purposes here is that a good strategy always has a clear time frame. In fact, it is the first element in articulating a complete strategy. Selecting, or determining, the appropriate time frame of analysis establishes the relevant factors to be included in the other elements.

As a strategy consultant. I have developed a generic strategy framework to help my clients articulate and evaluate their strategies. I have used some version of this model in multiple settings, but it was published in the book *The Sustainable Enterprise Fieldbook,* edited by Jeana Wirtenberg, Ph.D., William G. Russell, and David Lipsky, Ph.D.

Generic Strategy Model

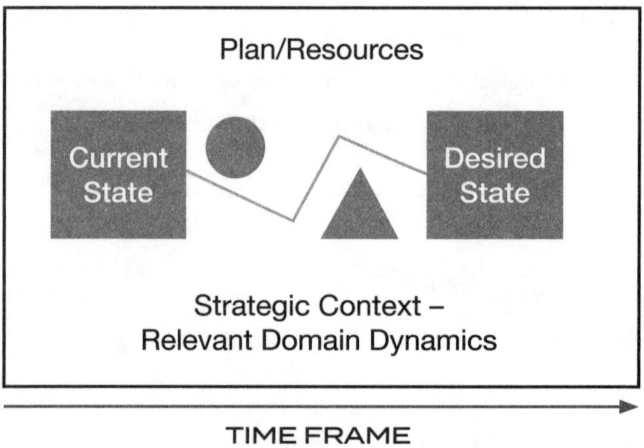

Generic strategy model elements

- Strategic Timeframe
- Strategic Context – Insight and Foresight into relevant domain and dynamics
- Understanding of Current State and Trends
- Vision of Future State
- Plan and Resource Requirements

Without going through this entire framework, I want to point out that determination of time frame for the proposed strategy is the first step. Establishing the time frame sets the frame of reference to determine what other factors are relevant.

TQ skills for identifying and adapting strategic time frames has significant implications for strategic leadership. Setting the strategic time frame determines what is relevant to the strategy formulation and development.

Although I have found this model to be effective in many situations, one particular experience always stands out. I was working with a large IT organization on developing a new set of innovations to grab a niche in a booming market. They did not have much time for consultants, but a previous client opened the door for a half-day executive team workshop to hash out the new product development strategy.

After a brief kickoff, I showed the generic strategy model and asked each of them to complete the model with their recommendation on the strategic approach. After thirty minutes, I moved them back into the group setting and asked for the work on the proposed strategy. What makes this stand out in my mind so vividly is that each one of them was using different time frames. This led to an almost complete mismatch in how they described the other elements. They were trying to get a common agreement on a strategy for solving different problems with different time frames. No wonder they were confused.

Nested Frames of Reference

An interesting aspect of time frames is that they are nested. Shorter time frames make up longer time frames and vice versa. The simple calendar is a good example. A day is a part of a week, which is a part of a month, which is part of a year. They are nested time frames.

Another way to think about it is the example of an odometer. Each tumbler of the odometer in your car spins at different rates. When miles add up to ten, the next tumbler rolls over one. When it gets to the next level for base 10, it turns over the third tumbler as hundreds—and so on. Anytime we are driving, all of the tumblers on our odometer are moving, just at different rates. Understanding that they are all nested in our total mileage is like recognizing the nested nature of time frames.

Time horizons are nested, so a one-year plan is a part of a two-year plan, is a part of a five-year plan, and the ability to focus on the right level of detail is the key to planning for the right level of detail for each. Likewise, a one-year plan can be made up of four quarters, and each quarter is made up of three months.

A good example of this concept is the nested time frames for a fan following their sports team. When asked, "How is your team doing?" the answer depends upon the nested time frame used to answer. We can pay attention to and assess any one play, which is part of each game. Each game is nested as a part of the season. And the season is part of the longer-term assessment of team performance.

Set/Subset/Superset

This concept is another triplet. One leads to another. Often our operating triplets are subconscious, and we develop mental habits.

The key for the second level of TQ development is to intentionally shift among multiple nested time frames. Everything has a logical three-tier dynamic: set, subset, superset structure. For example, an annual plan can be broken down into shorter times, or a quarterly plan. The annual plan can also be seen as a part of the longer-term three-year plan.

Time frame Reframe

Reframing your own time frame of analysis is a powerful way to enhance your understanding of the problem under analysis. Understanding is a function of context and when we change context, we change the meaning and implications of events.

There are a variety of methods for using a time-frame reframe, but they essentially involve intentionally shifting to an alternative time frame to analyze the problem from a different frame of reference. This can be done by shifting one's conscious focus to three perspectives including the nested time frames in the set, subset, and superset of the problem under analysis. This is the foundation for the Leadership Practice that follows.

Leadership Practice: Tri-frame nested reframe

Description:

One of the best ways to bring the time frame of analysis into conscious focus is to use a tri-frame nesting practice. This practice involves taking three different time frame perspectives on a specific problem under analysis.

This is a technique to shift the time frame of analysis being used to both shorter and longer time frames. Every specific situation has a specific time frame. Each time frame is a subset of a longer time frame and a superset of lower-level time frames.

Process Steps:

Using the tri-frame nesting practice involves three steps:

Step 1 – Recognize and label the time frame of analysis being used to address a particular situation or problem.

Step 2 – Shorten the time frame of analysis to identify and label the subsets, or component time frames, of the current plan.

Step 3 – Extend the time frame of analysis to identify and label the superset, of which the current plan is a component.

Practice Opportunity Triggers:

When delegating a particular task to someone, it can be used to remember to consider multiple time frames. Delegation is a leadership skill fundamentally dependent upon clarity of the time frame. Unless there's clarity of the time frame, it is not a clear assignment of responsibility. It is difficult to make sure that the expectations are understood.

Leaders must constantly evaluate the performance of others. This can serve as a trigger to reflect on time frames. Evaluation of performance always includes a time horizon and covers a certain period. Performance can be judged through different levels of expectation. We actually evaluate from multiple levels or multiple frames with different time frames.

I received the following comments in an email from a relatively new client:

"Thank you for giving me the gift of the tri-frame reframe. I had not realized it but I have often done this subconsciously. Breaking a problem down into subsets or remembering that every problem is a part of a bigger problem. However, in the past it had been unconscious, and I was not able to use that practice very well in communications with my boss.

"Once I grasped how important it was to get agreement on the nested priorities, I was expecting, I cut way down on lost motion. It also helped me accomplish key milestones on the path of a multi-year new product development strategy. Now I always ask, 'Which logical tier of the problem, short, medium, or long-term, are we focusing on at this time?'"

TQ Organizational Development Implications

Adaptability and agility to move between time frames

As we discussed in the last chapter, Jaques asserts that each individual has a maximum cognitive complexity/time horizon capacity at any stage of their maturation. This maximum determines their limits on the amount of complexity and the longest time horizon that a person can construct.

Although this is important in itself, it sometimes misses the point that most of the time leaders are dealing with the problems and opportunities that take place within their maximum capacity. A leader needs to be able to adapt different time frames to handle different problems. (Fig. 5.)

Time-frame adaptability and agility are the ability to intentionally shift to shorter and longer operating time frames of analysis. It is critical that a leader understand the dynamics of the subsets in the levels of work below their own. The problem arises when the complexity of the system or process under analysis requires a longer time horizon than they are capable of.

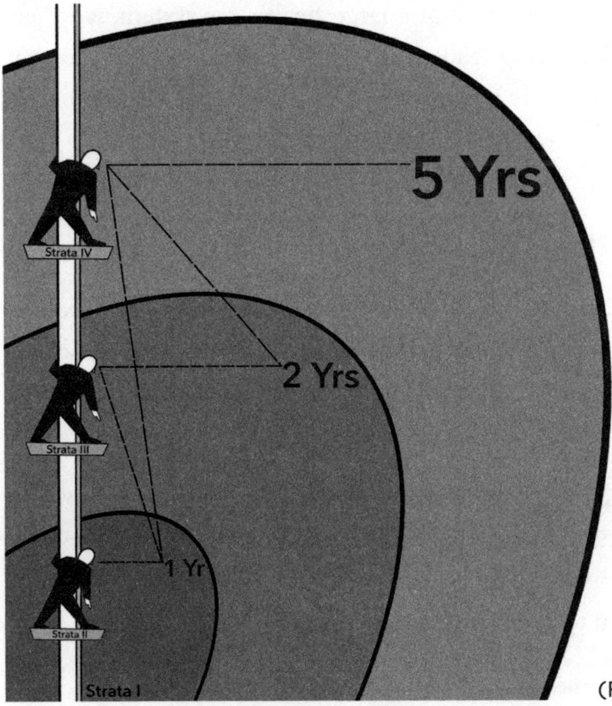

(Fig. 5)

In the case study that began this section, Craig Mack intuitively recognized this issue and consciously broke down the problem of re-engineering his firm into time frames that matched the capacity of his team. The key learning for Craig was that any strategy had to be executed using the time frames of his team, no matter how limited. Individual time frames are either a constraint or potentially an opportunity in any organization. It was his job to take his time frame and break it down for the organization and individuals into accessible, shorter chunks, organizing around existing time-frame constraints.

The inverse of managing to time-frame constraints is identifying individuals whose time frames exceed the requirements of their jobs, or potentially the organization as a whole. Under-utilized individual time-frame capacity is organizational gold because the individuals in question have the capacity to drive innovation, growth and value. It is one of the most underutilized resources in many organizations and perhaps the most compelling argument for TQ. Rich has a client, the CEO of a large healthcare system, who has made an art form of identifying leaders who have the capacity to develop billion-dollar business units. He is a master of TQ and its application to the fundamental business requirement for constant change and growth.

Reflections and Actions

- Since awareness of one's own time frame is a prerequisite to developing the ability to shift to multiple time frames, do you think these TQ skills are reinforcing our divergent? Can you think of ways to use these two dimensions of TQ together?
- How do you understand the concept of time-frame agility? How do you rate your time-frame agility? What can you do to improve your ability to intentionally shift the temporal frame of reference used to analyze a situation to alternative time frames?
- The tri-frame reframe is a simple but powerful practice. On reflection, do you find yourself doing this multilevel framing of the challenges in your work? Do you see when it might be a valuable practice? Make it a habit to always look at three levels of a problem when doing the creative work of problem analysis. Clarity among different levels is critical to good communication with others.

CHAPTER SIX

Awareness of Others' Time Frames

How do we understand what's going on inside someone else's head? It is hard enough for us to step back from our own operating perspective in order to have a different point of view and time frame. Consciously shifting your attention to explore another person's frame of reference requires an inquiring mindset. We can never really know what is going on inside someone else's mind, but we can ask questions and get a good idea about the time frame someone else is using in any specific situation.

Moving to the third level of development for TQ is a big step. The first two levels require introspection and adaptability of our own time frames and perspectives. After developing awareness of our own time frames, now our attention is shifting to focus on noticing the time frame of others.

Moving from awareness of one's own time frames to awareness of other's times frames is a big step.

The next two levels of TQ require shifting your point of view to consciously explore the temporal frame of reference of another person. Recognizing and connecting with the time frame being used by others involves asking questions and exploring their perspectives on the problem being addressed. This chapter addresses TQ Concepts and Leadership Practices for this third level of TQ.

Dimensions of Temporal Intelligence (TQ)

Level of TQ Development	TQ Description	TQ Leadership Practice
Self-Awareness	Aware of one's own *time frame(s)*	Meta-Perspective

Thesis: This third level of TQ is about noticing another person's time frame of analysis. Noticing something, anything, requires us to focus our attention and sort for the "thing" we are looking for. We only notice things, or patterns, when we pay attention and focus. Increasing awareness of others' time frames requires the interplay between reflective observations and inquiring questions.

Most of the time we all think our point of view is the rational perspective and that everybody else sees reality the same way we do. The way we understand the world and the frame of reference we use to determine what the relevant time frame should be in any situation requires mental discipline. The first step is to intentionally take a meta-perspective on your own frame of reference and time frame. The second step is to be curious about the frame of reference being used by the other person. Your mindset should be set on "inquiring mode." Asking good questions and looking for clues to the unspoken, always-present time frame of analysis, is the key to increasing the capacity to be aware of others' time frames.

TQ Concepts covered in this chapter:

- Expectations reflect time frames
- The meaning of "short-term"
- Needs/Outcomes/Impacts

Leadership Practice: Meta-Inquiry

• • • • •

Case Study Scenario

Continuing our discussion of his time frame as opposed to his partners', I said to Craig,

"You mentioned you're having a difficult time agreeing on these things. What do you suppose your partners are using for time frames? What's their context for all of this?"

Craig thought for a moment and admitted that he didn't really know.

"Have you ever asked them?" I asked.

"No. We keep talking about what must be done all at once, but we've never talked about what you're calling a time frame and what that looks like to each of us."

"Well, the reality is that you do have to address a series of interlocking problems at the same time—all the time. Each problem you address is contained in a relevant operating time frame. It is important to get insight into the way others are framing things and the time pressure they are responding to."

Craig's eyes lit up. "I know what you are talking about. Sometimes I find that the root cause of confusion is that different people are using different time frames of reference and therefore are looking at the problem differently.

"How can TQ help me be more insightful into the time frames being used by other people?"

"Let's explore a few examples."

Expectations reflect time frames

Exploring expectations is a very effective technique for understanding other people's operating time frame of analysis. All expectations have some sort of time frame. In general, we are unaware of the time frames others are using; and too often, so are they. As discussed earlier, time frames are normally operating in our subconscious mind, and we are not consciously aware of them. Probing and discussing specific expectations can help bring the operating time frame into view for each of us.

Good questions concerning expectations, with the intent to explore another's time frame, provides an effective way to better understand where someone else is coming from. Expectations, like time frames, are nested. One milestone is a part of a bigger process. It is also made up of smaller steps during the cycle.

The meaning of 'short-term'

What do you mean by short-term? Just asking this simple question can be a million dollars' worth of insight into someone else's perspective. It is shocking how many times the time frame I was using for "short term" was very different than how others on my team used that term.

For some people in specific situations, "short term" can mean today, or this week. In other cases, short term can mean the next two to three years. There is not a "right" length to be considered short term; it is always relative and subjective. Of course, the same logic applies to long term. "Long term" is only relevant when compared to what would be considered short term, or medium-term.

The point here is that we tend to use these familiar terms and think we have been understood. Or someone else uses the term and we mistakenly think we understand how long in calendar time they are referring to.

Early in my consulting career I had a client that used the term "short-term" to refer to that day and a month for "long-term". When I asked about the broader time frame within which the month nested, he seemed confused. He replied, "How can I think about what might come next if we do not hit our numbers this month?" Well...it turns out we had a similar conversation for several months before I realized that I was not the right coach for him.

However, in conversations with his boss I better understood. His boss could not, would not, focus beyond the quarterly numbers. The CEO that had brought me in soon fired them both—even though they had consistently hit their "short-term numbers".

Needs/Outcomes/Impacts

Another key TQ concept related to increasing our awareness of others' time frames is to distinguish between needs and outcomes. A "need" is perceived as a hurdle to overcome. An outcome is the desired objective. Needs tend to be framed as problems; as something that must be overcome. By focusing on the obstacle, or need, in the short term, we often lose sight of the desired outcome—which is always in a longer-term time frame.

Often there are multiple possible ways to accomplish the same outcome. Distinguishing between "needs" as intermediate short/medium-term milestones and "outcomes" as the longer-term results is an effective window for exploring someone else's operating time frame.

Outcomes are specific deliverables, or results, with a specific time frame. Outcomes are only as valuable as their impact. In most problem-solving scenarios, the desired outcomes are assumed to result in the real impact desired. As an example, you can provide a series of lessons to an employee, but they may not improve their performance as a result.

In general, needs have shorter time frames than do the associated desired outcomes, which lead to impacts rippling in all directions. People often cannot really focus on the longer-term outcome when they are paying attention to shorter-term hurdles addressing urgent needs. We often get so caught up like the proverbial metaphor, "so busy fighting alligators, we forget about draining the swamp."

Craig was trying to explain, *"My sales VP thinks we can get a quick boost in sales by paying higher commissions. She keeps saying we can do that tomorrow, but selling harder won't accomplish much without more sophisticated services. My*

Engineering VP says we should focus on hiring more engineers to work with the clients, but that would take at least six months to a year to make a difference and their capabilities will be diluted if the sales force can't accurately represent their capabilities and performance.

"On top of all of that, any improvements in service and sales won't make any difference to our bottom line unless our costs are lower. My Ops VP says we need to invest in a new facility with more modern equipment to improve our productivity and lower our costs. That's at least a three-year project. So, we keep going around and around on where to start."

I brought him back to our earlier conversation, "You have done the first thing. Now that you have locked in on your five-year time frame of analysis, the challenge is to understand the time frames of other key members of the leadership team to integrate the interdependences and overlapping time implications.

Leadership Practice: Meta-Inquiry

Description:

Meta-Inquiry is a technique used to probe someone else's understanding. Like the meta-perspective technique discussed in Chapter One, taking a meta-inquiry involves stepping back to a different level of observation. In addition to shifting to a higher-level perspective on your own time frame, exploring another person's time frame requires an inquiring and open mind. Meta-Inquiry involves recognizing that you need more information to understand the other person's perspective. What you are trying to discover is the frame of reference used by the other person to help them make sense of the situation. It is like connecting the dots to clarify the time frame being used.

Using probing questions concerning expectations and problem-solving is an excellent method of assessing other people's time frames. Of course, we never really know what someone else is thinking, so we must always approach it as an exploration, an approximation. It is not necessary to understand everything in someone's framework of analysis to determine their operating time frame.

Process Steps:

Step 1 – Recognize when you and another person seem to be out of sync in addressing a specific challenge.

Step 2 – Trigger yourself to use the meta-inquiry technique to explore their time frame of analysis.

Step 3 – Ask for clarity…

- on the timing of expected outcomes
- on the specific obstacles that need to be overcome
- on critical milestones required to accomplish the expected outcomes in a timely fashion
- on any element of future expectations.

Step 4 – Confirm time frame assumptions by asking for agreement on the timing of expectations.

Practice Opportunity Triggers:

In some of the core leadership functions, awareness of another person's time frame of analysis is an essential component to achieve capability for high performance. These include:

Planning—*all* planning—takes place in a future time horizon. Those time horizons are nested, so a one-year plan is a part of a two-year plan as a part of a five-year plan. And, the ability to focus on the right level of detail is the key to planning the right level of detail for each.

Time frame is also key to regular communication. When we are communicating with others our subconscious and unspoken time frames often lead to miscommunication. Remember to consistently ask, "WTF" (What Time Frame).

Delegation is another central leadership skill fundamentally dependent upon clarity of time frame. Unless there's clarity of time frame, it is not a clear assignment of responsibility. It is difficult to make sure that the expectations are understood.

Evaluation of performance always includes a time horizon. Performance is to be judged through different levels of expectation, which always include a time frame. When someone does not meet expectations, it is often because of confusion around time frames. We often evaluate from our own time frames and are not aware of others' time frames, again because of lack of communication.

The meta-inquiry practice is deceptively simple but sometimes it is difficult to keep an open mind of inquiry instead of reacting to the specific content. When I first began sharing the meta-inquiry technique many years ago, a client of mine pointed out to me, *"You made it sound so easy to just ask the questions to get insight into the time frame someone else is using. What I found out is that it is hard to ask good questions and keep an open mind.*

"When I would ask someone about their time frame, I was usually greeted with a blank look. Most people are not aware of the time frame they are using. I needed to learn to ask better questions instead of asking like a legal deposition. When I asked questions about when key events or deliverables were expected, their responses reflected their time frames. So, I learned to ask good questions that required someone to reveal the contextual time frame of reference."

TQ Organizational Development Implications

Organizational levels and time frames of expectation

According to Elliott Jaques and his Stratified Systems Theory (SST), organizational levels should reflect the complexity of the tasks required in the time frame of responsibility. Gaining insight into the time frame someone else is using does not mean that you should be using that same time frame.

Different roles at different levels of an organization have different time frames of responsibility.

In the next chapter, we will discuss the challenge of getting in sync with others in terms of the time frames for different processes and will be referring back to the chart below. For now, the point is that organizational levels should have different time frames of responsibility. A chart from Jaques's "Requisite Organization" showing the organizational strata for different time horizons follows (Fig. 6.)

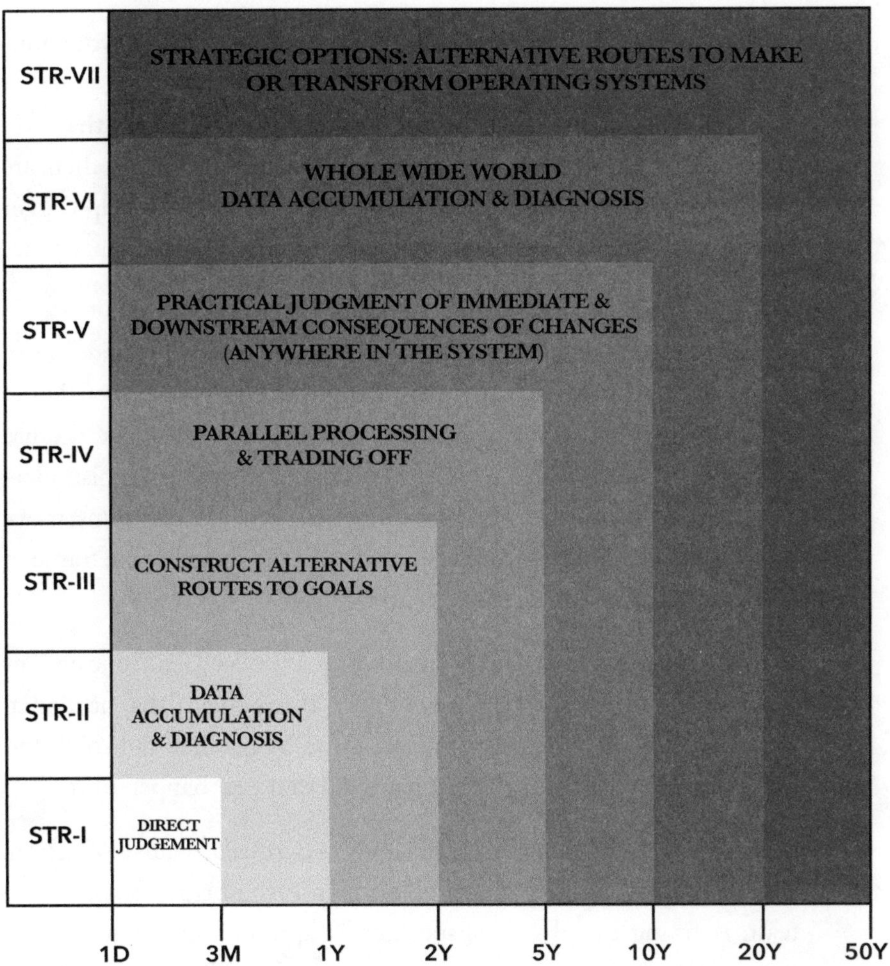

(Fig. 6)

Elliott Jaques uses the term *time frame of intention* to mean the length of time required to meet the expectation of a specific delegated task. He acknowledges how we use different time frames for problems of different complexities. The challenge is to be flexible enough so we can observe shifts in a person's time frame of analysis. We do not want to lock in our observation of one's time frame and miss the often-rapid shift to a shorter or longer time frame.

Reflections and Actions

- On a scale of 1-10, 10 being the highest, how do you assess your ability to notice the time frame others are using when working together in a problem-solving mode? Have you ever thought about that skill before? Can you see how often you make assumptions that others are using your time frame of reference? What can you do to help trigger your shifting to a meta-perspective and using the meta-inquiry practice?

- Have you ever been in a situation where people agreed on "short-term action items," but were not clear on when things would actually get done? What was the outcome? There are many cases where the use of "short-term" and "long-term" have caused significant misunderstandings. In the future, when someone uses one of those terms, move into a meta-inquiry mode and get clarity on the specific time frame of expectations.

- Can you recall a time when you realized that the "problem" one of your team members was having was not really "the problem" that needed to be solved? People often get attached to problems they can solve and ignore the broader strategic context. Does that ever happen to you?

- A "problem" is often a barrier blocking a particular course of action, but simply overcoming that problem is not the bigger problem the team is trying to solve. Yes, specific disruptions can require urgent attention, but they are only steps in a longer plan. Make it a practice to get clarity on *which* problem is under analysis.

CHAPTER SEVEN

Ability to Sync with Others' Time Frames

What do we mean by "getting in sync"? Getting in sync is a buzzword for synchronizing perspectives in terms of analyzing a problem using agreed-upon time frames and expectations. Most people do this naturally. Good communicators often are in sync with the others in terms of the way they frame a situation. We feel understood when we are using a common time frame and context.

Synergy in teamwork comes from getting different perspectives aligned with a common time frame.

Developing our ability to shift our own time frames provides the foundation for helping others shift theirs. The previous level of TQ focused on intentionally noticing another person's operating time frame. Level Four of the TQ development model is about getting in sync with another's frame of reference and creating a common understanding of the situation or problem under analysis.

Level Four TQ is built on the three earlier levels. This fourth level is focused on the ability to get in sync with others' time frames. First, we must recognize our own time frames and be able to adapt those to multiple levels. In addition, we need to be able to recognize the time frame another person is using to understand a problem. This fourth level builds on all three lower levels and provides a foundation to connect around time frames to be used to address a specific situation or problem.

Level of TQ Development	TQ Description	TQ Leadership Practice
Self-Awareness	Aware of one's own *time frame(s)*	Meta-Perspective
Ability to adapt own	Adapt own *time frame(s)* of analysis	Tri-frame Reframe
Awareness of Others	Aware of others' *time frame(s)*	Meta-Inquiry
Ability to Sync with Others	Help others adapt their tasks and *time frame(s)* of analysis	Tri-level Reframe

Thesis: The central concept of this chapter focuses on one's ability to get in sync with the time frames of others. As mentioned above, getting in sync, or getting in alignment, includes both having awareness of our own time frames and also being able to lead other people to share a common time horizon for specific expectations.

Agility is the description used because harmonizing time frames requires balancing at least two different perspectives. When time frames are aligned, it often feels like there is a "common understanding" of the situation and alternatives.

At times, it can be a challenge because more than one person is involved in a problem-solving analysis. In this case, it is necessary to recognize and align with multiple time horizons simultaneously.

TQ Concepts

- Exploring nested time frames
- Value of a common timeline
- Purpose/Perspective/Perception

Leadership Practice: Tri-level Reframe

• • • • •

Case Study Scenario

I asked Craig to think about how he could explore his partners' time frames. Initially, he said that he could just ask them. I responded that it was a good place to start, but then I asked him how his sense of responsibility to his father influenced his longest time frame—his five-year goal. I wanted him to think about how what he needed influenced his intentions.

"That's the most important thing to me and everything else is about achieving that."

"Okay, I said. "What's the equivalent for each of your partners?"

We spent a few minutes talking about the three executives he mentioned earlier, and it turned out that he thought he had a pretty good grasp of what was important to them. He was surprised that he had never connected their perspectives with different time frames.

Then I asked, "How do you use that insight to get everybody in sync on expected outcomes within specific time windows?"

After some thought, he said, "The different initiatives are overlapping in time, and it needs to be clear how they work together to get us where we need to be in five years. It is not an either/or. To get where we need to be we have to do them at the same time. Like walking and chewing gum," he laughed.

"How does TQ help me get in sync with my partners?"

"That is exactly what we need to talk about," I replied. "Let me suggest a few concepts and practices that can help develop those temporal skills."

We then set up a schedule to work with and develop TQ skills at all four levels.

TQ Concepts

Exploring nested time frames

As has been discussed several times in earlier sections, time frames can be understood as being nested, each larger time horizon including the previous. Chapter Two discussed being aware of our own nested time frames. At this TQ level, we are exploring the temporal perspective of others by creating three nested time horizons when analyzing a specific problem.

To develop our ability to recognize and get in sync with other people's time frames, we can ask questions to explore their perspective by asking good questions. Exploring the nested nature of time frames helps make the connection between the short term, the medium term, and long-term objectives and how those work together for better outcomes.

In an ideal situation, people are clear about how the subsets roll up into broader and more complex sets of issues and that those sets are a subset of an even larger set.

The meta-inquiry technique introduced in the last chapter is also an important element for developing TQ Level Four. Asking open-ended questions about purpose and perspective allows us to assess the time frame being used by the other person in this case. When we are exploring several different and interconnected purposes and perspectives, finding the intersections and overlaps is a good way to get into sync.

Value of a Common Timeline

One of the first things I learned in my consulting career was the value of having a common understanding of the timeline that everybody is working from. Detailed work plans are great tools but can often get bogged down in a level of detail that masks the basic timeline. A timeline is a simple linear projection of when activities will occur, so that key objectives will be accomplished.

I quickly learned that the senior partners in my firm had a very different understanding of the time frame of engagement. I was working on a 90-day plan with challenging deliverables and limited access to key resources. When I invited myself to sit in on a project leadership meeting, I was shocked to realize my chunk of the work did not even show up on the timeline. The plan was laid out over five years and was at a completely different vantage point than I had been using.

This experience was merely a crack in the wall. I realized that this is a common problem when I cannot see my part in the bigger picture. Developing a commonly understood timeline can be an easy way to get people from different levels and functions to understand where and when they fit in the plan.

Purpose/Perspective/Perception

This is another triplet—one leading to the next. Purpose is a person's intentional outcome at some point in the future. This purpose then determines one's perspective, or point of view. In turn, perspective creates the filters for what we notice and what we do not pay attention to.

Like time frames, our purposes are nested. We have long-term aspirations as well as nested medium- and short-term hopes and goals. Also, we have multiple purposes associated with a range of roles, ranging from work to home to civic, social, and recreational objectives and priorities.

Each role has its own nested set of responsibilities for different people at different times. Clarity of purpose is a key to high performance teams. Purpose creates the relevant context, or perspective, to understand and make sense of a situation or problem.

Clarity of purpose is a key to high performance teams.

Our purposes determine our perspectives. How we see the world is driven by what we're trying to accomplish at multiple levels. Perspective, as we've been discussing, always includes a time frame. Time horizons and other elements of our subconscious context determine our perceptions, or what we notice as the relevant factors in a situation.

Getting in sync with another person involves alignment around a common purpose, perspective, and perception. All of these mental frames have temporal dimensions and time frames. Clarity and alignment around purpose requires clear framing to achieve a common perspective and to identify the key factors to be noticed.

Purpose determines perspective which filters what is perceived.

In my early work with the Quality Control movement, I learned a treasure trove of concepts around how making something a priority and focusing on it will lead to better results. In fact, that is essentially what quality control is—identifying key data points that can be monitored to provide insight into common mistakes. I found it amazing that most people do not have a clear timeline for the processes they support. A simple, commonly understood timeline was the key to designing an effective system of checks and balances. If we were not in agreement on "when" something would happen, it was impossible to agree on the "who" and "how".

Leadership Practice: Tri-level Reframe

Description:

One of the best ways to focus your awareness on another person's time-frame of analysis is to use a tri-level reframe practice. This is a technique to observe another person's timeframe of analysis by shifting to both shorter and longer time frames. Every specific situation has a specific time frame; however, that time frame is a subset of a longer time frame and a superset of lower-level time frames.

Process Steps:

Using the Tri-level Reframe involves five steps:

Step 1 – Ask questions about the specific time frames on what they consider short-, medium-, and long-term factors relevant to the issue under analysis.

Step 2 – Recognize and label the time frame of analysis being used to address a particular situation or problem.

Step 3 – Shorten the time frame of analysis to identify and label the subsets, or component time frames, of the current plan.

Step 4 – Extend the time frame of analysis to identify and label the superset, of which the current plan is a component.

Step 5 – The objective is to notice how they frame the situation into three nested time frames. Once the time frame of analysis being used by another person is determined, we are able to move to the next level of TQ and the next chapter, which focuses on our ability to sync time frames with others.

Practice Opportunity Triggers:

- **Delegation** is another central leadership skill fundamentally dependent upon clarity of time frame. Unless there's clarity of time frame, there is no clear assignment of responsibility. It is difficult to make sure that the expectations are understood.

- **Evaluation** of performance always includes a time horizon and covers a certain period of time. Performance can be judged through different levels of expectation. Actually, we evaluate from multiple levels or multiple frames with different time frames.

A student in one of my workshops made the following comments about using the Tri-level Reframe technique: *"From the beginning, this technique had an intuitive appeal to me. I could see how alignment of sets, subsets and supersets would solve some of my problems with my boss and my boss's boss.*

"When I tried to suggest the technique in a senior staff meeting, I had surprising results. My recommendation for how to break the project into nested sets of activities was immediately rejected. However, we spent the next thirty minutes discussing the set/subset/superset framework from the big boss's perspective. At the end of the day, it was not the way *I broke it down. It was the value of breaking it down as a* team *that was the real value."*

TQ Organizational Development Implications

Quintave (five-level) model of mental process

Jaques leveraged the work of Piaget and other development psychologists to develop a Quintave Model of Mental Processes. Within each level of cognitive capacity are four distinctly different types of reasoning. This concept was introduced earlier and will be fleshed out a bit more here.

Again, a Quintave is a repeating set of four types of logical processing with the first repeating as the last. Like an octave in music, there are eight notes, but the first and last is a C. They repeat up and down the scale. In a similar way, the levels of abstraction associated with each type of processing are distinct and repeat at higher levels of complexity (Fig. 7).

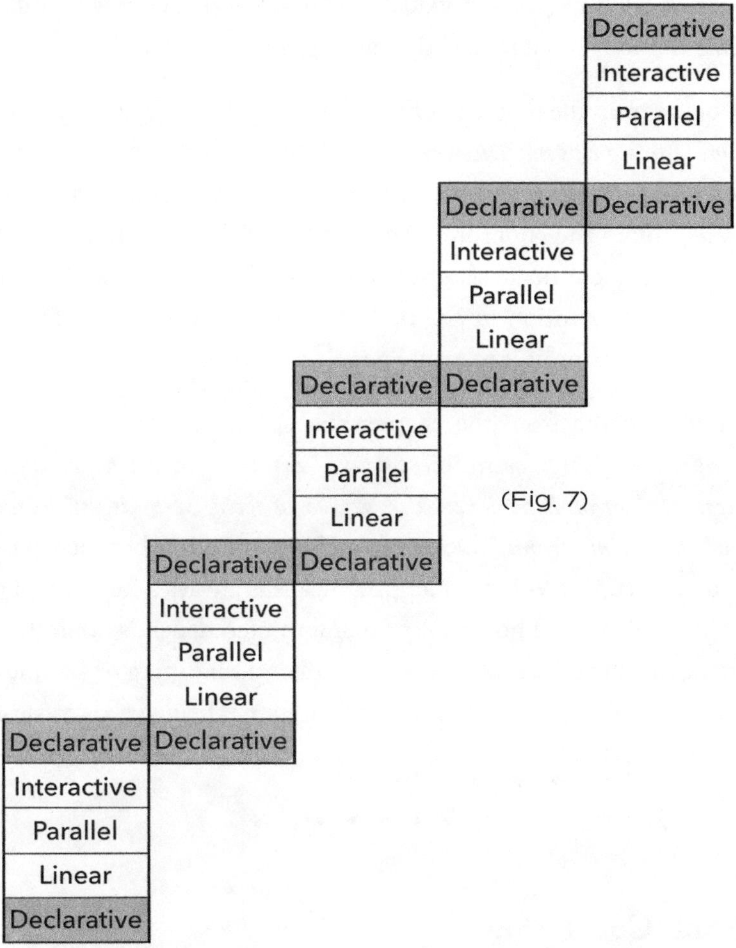

(Fig. 7)

Reflections and Actions

- Do you find it natural to adapt your time frames to get in sync with others? Or do you usually work with the other person to get them to see things from your point of view? In either case, getting in sync on time frames of analysis is critical. Reflect back on times when this problem erupted into much bigger consequences? What could have been done better to prevent the misunderstandings that occurred because of different individuals or groups operating with different time frames? What is an early warning sign of this kind of confusion?

- A quote from the first key concept states, *"Exploring the nested nature of time frames helps make the connection between the short term, the medium term, and long-term objectives and how those work together for better outcomes."* Does the concept of exploring another's time frame to get in sync with them make sense to you? Do you do that now? How well? Create the intention to practice with the tri-level reframe technique early in a new project to assure people are in sync.

- Another quote from the chapter states, *"Our purposes determine our perspectives. How we see the world is driven by what we're trying to accomplish at multiple levels. Perspective, as we've been discussing, always includes a time frame."* Do you see the connection between purpose and perspective? What something means is always a function of one's point of view. And how things are interpreted depends upon the time frame of analysis. When there are conflicts in meaning or significance, it is often when there are different time frames being used to assess the potential impact.

• • • • •

Case Study Conclusion

Following up on our case study with Craig and his challenge to align different initiatives within a common time frame, TQ played a valuable role. It was not easy at first. Each department leader had good logic and a solid plan for what needed to be done in their respective areas. The problem was that they could not all fit within the time and budget available to meet the five-year time frame for payout required to meet his commitment to his father.

There were a range of things that needed to be done to assure the firms competitiveness and profitability and each of them was "critical". But the reality was that they could not all be done in the same time frame. He had to get them aligned.

By establishing the five-year payout to his father as the "strategic time frame", he was able to work with the different groups within the firm to adapt their initiatives to fit within that time frame. In some cases, there were investments that just had to be postponed. In several others he was able to get satisfactory results without getting ideal results. In two situations, refining the plans within the "given time frame," resulted in major gains without nearly the investment in time and money originally proposed.

Craig was pleased. His father was pleased. And all seemed well to them—until he had to start grappling with the next five-year plan. But that is another story.

PART THREE

Cycle Timing

INTRODUCTION

What are the critical processes and cycles you are responsible for in your leadership role? How long do they take to complete? All processes, especially business processes, are designed and built with an outcome expectation. Developing awareness of the timing of repetitive cycles is an important TQ capacity.

The previous part of this book focused on linear time measured in time frames. These chapters focus on cycle-time and rhythm. Understanding and developing a sense of timing is working with **Kairos Time**, or cycle time. As discussed earlier, the Greeks had different words for different types of time. Linear time was called Chronos, while cycles and rhythm were called Kairos time.

Thesis: Developing awareness of the timing patterns of processes and cycles is an essential leadership capability. Completion of a process marks the end of a sequence of activities, leading to the final event, which marks the beginning of the same sequence of activities in the next cycle. The length of time required for things to happen is "cycle time."

Our sense of timing is both cognitive and intuitive. Timing is a skill that can be developed. Our cognitive development in this area is based upon improving our ability to notice key events and process milestones that determine when a cycle is coming to an end and another one begins

The central concept of this chapter is that everyone can improve their sense of timing by intentionally paying attention to cycle time and cycle-timing. Based upon noticing how long different processes and cycles take to complete, we can adapt our mental algorithms and maps.

We use Kairos time to estimate when a process is going to come to completion. Examples of classic cycles in life include term of pregnancy, seasons change, stages of life. In a leadership role in an organization, classic cycles include the sales cycles, production cycles, and accounting cycles. The point of TQ is that we need to be more aware of how we estimate and facilitate key events, processes, and cycles.

TQ also addresses the question of how we estimate the time a process will take to complete. We all have mental "algorithms" that yield our sense of timing. Forecasting expected outcomes is essential to leadership. And, like other TQ skills, we can improve our sense of timing.

• • • • •

Case Study

In this final section we are going to work with a good case example of how understanding and alignment of cycle times came into clear focus.

Mark Patterson, the client, was a new business unit president for a fast-growing MedTech company on the West Coast. They were in a hyper-competitive market with new technologies disrupting the sector. Staying on the bleeding edge of the new technology curve was imperative, but this required a high level of integration among multiple key business processes.

A longtime colleague had recommended me to Mark, and I was engaged to help align "burn rate" with the timing of new cash tranches from the parent conglomerate. After I had done some research, we got together to discuss his situation and scope out my engagement. After a very brief introduction, he jumped right in. I did not have to prod him to talk.

Mark launched into a good summary of his situation, *"My new role is a big opportunity, but it could also be a suicide mission. The challenge is to stay just behind the "bleeding edge" with research and new technology development but be the first to market in our primary target sectors."*

The holding company had plenty of money to invest, but the why, what, and when questions had to be clearly answered and new tranches of funding were based upon key results, not definite dates. This makes clarity of integrating business processes essential to good planning and forecasting.

The basic challenge he described was all about accounting and finance. He seemed to be missing the point that business processes drive accounting, not the other way around.

He went on a bit but then stopped and asked, *"I have two questions. One, do you think you have a general understanding of my challenge? And two, how do you think you can help?"*

With confidence I looked him in the eye and said, "I'm not sure you really understand your challenge. You are describing your funding requirements for several different parts of your real problem. Your problem is not driven by funding cycles. It is driven by business cycles. You need to better understand the way the different departments of R&D, production, and marketing overlap and weave together to define the "how much by when" based on those cycles."

"Let's get started," he replied with relief.

Moving to the white board, I said, "Let's first discuss a few basic concepts to help think about the interdependent timing of different business cycles."

Overview

The three chapters in this third part of this book:

Chapter 8 – *Awareness of Cycle Time* discusses time as a cyclical process and how we can become more aware of the cycles we manage as leaders.

Chapter 9 – *The Sense of Timing* focuses on both our rational and intuitive senses of when things will happen. Timing relates to cycles of any length—from seconds, to days, to months, to decades. Everyone has a sense of timing, but it is usually subconscious and difficult to analyze or improve. Developing our sense of timing is an essential TQ skill.

Chapter 10 – *Improving Timing Skills* explores how we estimate timing, and how we can become better estimators of time requirements for critical business processes. Having a good sense of when those multiple nested cycles will evolve through their life cycles is what makes a high-performance leader and high-performance leadership teams.

Each chapter follows a similar structure with a discussion of **TQ Concepts, Leadership Practice, Organizational Development Implications**, and **Reflections & Actions**.

The foundation of this book is that learning interesting ideas is interesting. Learning to use those ideas in your leadership role is the real objective. And guess what, that takes time and effort.

CHAPTER EIGHT

Awareness of Cycle Time

The focus of this third part shifts from understanding time in linear terms to understanding time in terms of processes, cycles, and rhythm. This chapter addresses the key concept of "cycle time" and how it plays an important role in leadership.

Thesis: Developing awareness of the timing patterns of processes and cycles is an essential leadership capability. Completion of a process marks the end of a sequence of activities leading to the final event, which marks the beginning of the same sequence of activities in the next cycle. The length of time required for things to happen is "cycle-time".

Organizations, and organisms, are made up of multiple symbiotic subsystems. Systems are made up of numerous parallel and interdependent processes. Most business processes are repetitive and occur in regular cycles. Processes are linear ways of looking at cycles—they begin and end. Linear-time can be used to measure the length of time required to complete a process and deliver an outcome.

The time to complete one revolution of a cycle is cycle time. Noticing cycle time is a necessary first step for an individual to improve his or her ability to accurately estimate how long a cycle will take to complete. Paying close attention to cycle time is the basis for improving one's timing.

Developing awareness of the timing patterns of processes and cycles is an essential leadership capability.

The variety of cycles is never-ending. They each have some basic similarity, even when they have widely different content. Examples of business process cycles include market cycles, sales cycles, career life cycles, accounting cycles, technology cycles, new acquisition cycles, new product development cycles, new regulation cycles, and many more.

Some of these cycles are easy to observe and measure in time. In fact, systems like accounting are based upon date-driven cycles. However, the last three business processes listed above are examples of processes that can have a high degree of time variability. Cycle-time variability is often a function of interdependencies and resource allocation.

TQ Concepts covered in this chapter:

- Every "*thing*" has a life cycle
- Cycle-time mental algorithms
- Assumptions/Dependencies/Probabilities
- Events/Trends/Cycles
- Nested Cycles

Leadership Practice: Noticing one's own cycle-timing algorithms

• • • • •

Case Study Scenario

After the very first two-hour session with Mark, we both had gained a lot more clarity about which business cycles were most critical and needed more insight into how the different functional areas of the business had to coordinate.

Mark summarized our discussion saying, *"This has been very helpful. By thinking through the cycles and sub-cycles from different parts of the business, it really helps us understand the key drivers on the critical path. We are going to be making a lot of assumptions, but at least we can define those assumptions so that we can monitor them and adapt quickly when things turn out differently—and they always do."*

We then started sharing some of the TQ concepts with other senior team members and began documenting the different business processes driving the new product development and release schedule.

TQ Concepts

Every "*thing*" has a life cycle

The term "thing" is the most often used word in the English language. A "thing" is a generic generalization of any object. We can also call a thing an "it". Any noun and many verbs can be turned into objects or things. A *thing* is a recognizable pattern that occurs over time. It turns a dynamic into an object. Events are part of the processes that occur in time. Therefore, every "thing" is a process that has a beginning and an end. Some take nano seconds and others take decades or centuries. We recognize and label these processes as "things". The length of time between when a cycle begins until it comes to completion is the life cycle of that "thing".

Even physical things have life cycles. Every physical object had a beginning point in time and at some point in the future it will dissipate and cease to exist in the same form. In a similar way, abstract "things" like systems and processes have a beginning and an end. No things are permanent, but cycles repeat. TQ in this area starts with paying more attention to the life cycle of the systems and processes that are important in your work role.

Cycle-time mental algorithms

"Mental algorithms" provide a good metaphor for how we think about things and make decisions. The construct of mental algorithms has a great deal of utility when understanding the structure of mental processes, in contrast to the content associated with a specific process. Discussion of algorithms has become avant-garde these days as people try to embrace the concept of artificial intelligence based on algorithms. An algorithm is a logical calculation that includes feedback loops as a basis for adapting the logic model to include emerging trends. In this case, the term "intelligence" is being used to mean the ability to learn and adapt.

I think there is more to intelligence than that adaptive capacity. However, adaptive data algorithms provide intelligence to a decision-maker; the data produced is not "intelligent" in and of themselves. Now, the human intelligence of the decision-maker to effectively use the information, or intelligence provided by data, is another question.

Our normal mental processes can be thought of as equations. Much of the time we run repetitive programs, habits that repeat the same thoughts and activities when facing a similar situation. "This is another one of those." Instead of simply repeating the same equation and getting the same answer, intentional mental algorithms adapt based upon feedback loops that recognize emerging patterns. We will be discussing mental algorithms several other times in the following chapters.

Estimating when something will come to completion, or a key event will happen, uses our internal algorithms. The ability to observe and notice pattern changes and feed that back into our calculus of probability is how we refine our cycle-time algorithms. The better our algorithms become, the better we are at predicting the timing of outcomes or what we're calling "cycle timing".

Mental algorithms are usually subconscious. But we can pull them into our objective consciousness and analyze our thoughts and decision processes, but the actual operation of our mental algorithms is largely subconscious as well. Reaching into our subconscious to adapt our mental algorithms; to adapt our calculus with better feedback loops, is at the heart of temporal intelligence.

An example of this important point comes to mind. One of the direct reports of a coaching client always seemed to have "no problems until there was a problem." He was always optimistic about his team's performance, but he almost never got anything completed on time. My client was frustrated, and it turned out, so was the other guy.

I asked him about the key cycles that drove his performance and he did not have a clue where to start. So, we walked through his area of responsibility and the key business cycles that included and he went away to ponder.

About six weeks later, my client reported that he had dramatically improved his ability to meet deadlines. He shared that the key was having a clear enough description of the processes that make up the key cycles so that you can monitor progress, virtually day by day.

Sounds easier than it is, but it is doable.

Events/Trends/Cycles

This is another triplet. Algorithms are developed to track key events and integrate the data back into the decision-making processes. Algorithms are also designed to keep track of trends, which is very different than simply observing key events.

Monitoring trends is a classic algorithmic process. Predicting trends is what computer algorithms do best. Good leaders also use mental algorithms to understand the cycles and business processes.

Noticing cycles emerges with experience. Good leaders develop the ability to intuitively integrate emerging key events and dependencies that affect the timing of those cycles in their area of responsibility.

Assumptions/Dependencies/Probabilities

Another triplet.

Algorithms are built on **assumptions** and decision model criteria. Understanding internal algorithms is largely about understanding the assumptions built into our mental models.

Another key element of algorithm processing is recognizing that many of the variables have **dependencies**. Each of these interdependencies can have an effect on the time required to complete the cycle.

The way we construct the future is to extrapolate **probabilities** based on past experiences, beliefs, and assumptions. It is in the intricacies of these probability projections that cycle timing is most evident.

The case study with Mark provides a good metaphor here. Mark had to create a plan and budget that integrated the three primary business functions of R&D, Production, and Marketing. As his team and I dove into the details, it was clear that there were multiple places where the different functions had a "critical-path handoff".

Initially, they had diagrammed their key processes without clear linkages to the details of the other departments' timing. As they worked to understand their own processes and cycles better, they were better able to analyze their assumptions, dependencies, and probable time frames. As a result, each group developed an excellent visualization of their overlapping processes and, therefore, budget requirements and timing.

Nested Cycles

Since cycles can be measured in linear time, like time frames, they are nested. If you recall the discussion in Chapter 5, we discussed how time frames are nested—with shorter time frames occurring in longer time frames, which occur within even longer time frames. Part of the challenge of noticing cycles by observing key events and trends is distinguishing among the nested processes and cycles.

One good practice is to identify cycles that are relevant in the short term, those that are relevant in the medium term, and cycles that are relevant in the longer term. Of course, nested cycles can refer to any length of time.

For example, your commute has a cycle time. Your work week has a cycle time. Many other important cycles are unfolding over months, years, and sometime decades. Some of the relevant cycles in a business, such as new product development, often extend to five, ten, twenty years or more. Distinguishing among the interdependent cycles that occur within different time horizons is a valuable TQ capacity.

Distinguishing among the interdependent cycles that occur within different time horizons is a valuable TQ capacity.

Leadership Practice: Noticing one's own cycle-timing algorithms

Description:

This practice is based upon reflecting on your own thought processes and learning to pay attention to how effectively you estimate how long a cycle will take to complete. Like other practices provided in this book, the basic challenge is to break into your stream of consciousness to intentionally focus on your own mental thought processes.

Process Steps:

Step 1 – Set and use a trigger event to allow you to shift to a meta-perspective.

Step 2 – From this neutral space, focus your attention on the expected life cycle of the event or trend under analysis.

Step 3 – Then shift to notice how you came up with that estimate of cycle time. Bring to your consciousness the factors used to consider the most probable timing of the outcome.

Step 4 – Reflect on how you calculated the expected timing and consider what feedback might be incorporated in that particular "estimation program".

Step 5 – Set the intention to notice how accurate your time estimations are and consider adapting your mental algorithms.

Practice Opportunity Triggers:

The best trigger opportunities for noticing your own "estimation programs" is when you are doing planning. Planning is basically an exercise in estimating when things will happen. Too often we just find a convenient date and declare that is the completion date. It is far better to practice reflecting on your mental algorithms and building a logical framework for estimating.

In addition, when evaluating someone's work, this provides a trigger opportunity to better understand why things did not happen as planned. What can be learned so the miscalculation is revised in the future.

TQ Organizational Development Implications

Cycle time and organizational span of control

Different individuals operate with different time horizons based upon the nature of the work at that level. Each organizational level requires different time frames to manage the cycle time of the work. A leader's level in an organization should be within his or her cognitive capacity to operate within that time horizon. Each time horizon reflects the tiers of complexity of the work under that person's supervision.

The table on the following page (Fig. 8) depicts strata for organizations, leaders as complexity below depicting the classification approach for both organizations, leaders, and complexity.

Organizational Levels & Time Span of Role
Statified Systems Theory

Maximum Time-span of Role	Organizational Level or Strata	Industry Roles	Level of Task Complexity
50 Years	Strata VII	Holding Company CEO	Construct Complex Systems
20 Years	Strata VI	Holding Company EVP	Oversee Complex Systems
10 Years	Strata V	Business Unit President	Judge Downstream Consequences
5 Years	Strata IV	General Manager	Parallel Process Multiple Paths
2 Years	Strata III	Unit Manager	Create Alternate Pathways
1 Year	Strata II	First Line Manager	Diagnostics Accumulation
3 Months	Strata I	Operator/ Staff	Practical Judgement

(Fig. 8)

Based upon Jaques's research, leaders can be classified as being on one of seven levels based upon the time horizon they use to frame an opportunity or a problem. The cognitive capacity, and therefore time horizon capacity, of a leader needs to be sufficient for the level of his or her organizational role.

Reflections and Actions

- Have you thought about cycle time before this chapter? How well do you think you grasp the core notion? How would you explain it to your boss?
- In the first TQ concept it says, "A "thing" is a generic generalization of any object. We can also call a thing an "it". How does this concept grab you? Try noticing how often we use the term. To say any*thing*, about any*thing*, requires us to define some*thing* as a "*thing*".
- Are you aware of some of your current mental calculation programs? We have thousands of "mental shortcuts," or mental programs, that we run constantly. But does it make sense to think of these mental programs as algorithms?

- "An algorithm is a logical calculation that includes feedback loops as a basis for adapting to emerging trends. In this case intelligence is being used to mean the ability to learn and adapt."

  Does this make sense to you? Have you thought about it like that before? Can you see how this concept could be useful in refining your own TQ skills?

- Another quote from this chapter, "One good practice is to identify cycles that are relevant in the short term, those that are relevant in the medium term, and cycles that are relevant in the longer term."

  What are the short-, medium-, and long-term cycles that make up the key parts of your leadership role? How much time do you spend on shorter instead of longer time frames? Take a few minutes and identify the key cycles in your work that need to be monitored.

- The final key concept states, "Different individuals operate with different time horizons based upon the nature of the work at that level. Each organizational level requires different time frames to manage the cycle time of the work."

  Have you thought about how organizational hierarchy reflects the time span of the cycles each level is responsible to manage? How well is this reflected in your organizational structure? Examine your role and the time spans of the cycles you are responsible to manage. Identify key factors that need to be monitored.

CHAPTER NINE

The Sense of Timing

What is "timing"? How do you assess your current "sense of timing" relevant to your leadership role? This chapter covers an intriguing topic—our sense of timing. A good sense of timing is invaluable for effective leadership, but it is not a skill we either have or do not have. Like most skills, our sense of timing can be developed through intentional focus on the structure of the mental processes for how we notice and understand cycle time.

The previous chapter introduced the concept of cycle time. This chapter takes it a step further and examines the way we anticipate outcomes and ways we can improve our sense of timing.

Thesis: The thesis of this chapter is that our sense of timing is both cognitive and intuitive, and that timing is a skill that can be developed. Our capacity in this area is based on our ability to notice key events and to process milestones that determine when a cycle is coming to an end and another one begins.

Our sense of timing is both cognitive and intuitive.

The world is in a constant flow. Understanding how specific processes arc through their life cycle to reach completion is an essential leadership capability. The accuracy of our estimated time of completion is what determines good timing. Good timing is a critical capacity for a high-performance leader.

What are the key indicators for observing how quickly or slowly a process or cycle is unfolding? Which patterns should you notice and how do we come to understand the phases and stages that make up their life cycle? The challenge is to figure out which things to pay attention to, and when to notice.

TQ Concepts covered in this chapter:

- Timing includes an Intuitive Sense
- Opportunity Vision
- Expectation/Observation/Adaptation

Leadership Practice: Noticing others' sense of timing

• • • • •

Case Study Scenario

Again, reflecting back on our case study with Mark and his challenge to "lead the market with new technology." After a working session where each of the four teams—R&D, Production, Marketing and Finance—developed their own critical process mapping, the challenge for Mark and the team was to bring them together into a single plan that provided the new-cash road map.

After a brief overview by Mark, I gave the combined team thirty minutes to identify where their process might endanger the timing of other departments and groups and where their vulnerabilities were in terms of interdependent processes. What they came back with was amazing in two ways. One is that they all identified where they were vulnerable to the timing of other parallel processes, but none of them had adequately described how they created vulnerabilities for other departments.

Second, they demonstrated a wide range of forecast on some key issues. This led into a discussion on "cycle-timing".

TQ Concepts

Timing includes an Intuitive Sense

Timing includes an intuitive sense of flow. There are the mechanical, cognitive aspects of good forecasting, but there is also a non-logical sense that needs to be cultivated. We can consciously notice key events and trends and improve our sense of timing, but beyond all our mental mapping is an underlying intuitive sense of timing that is more than cognitive reasoning. Sense of timing is like rhythm and beat, sensed more than grasped.

We are multi-sensory beings with a wide range of "sensing". We are often confused that our five primary external senses are our only way of making sense out of our world. Just think about our sense of balance, our sense of humor, our sense of danger, our sense of right and wrong.

One of our key senses is our sense of timing. It operates from moment to moment, and it also operates over years and decades. Our sense of how long things will take to evolve is essentially intuitive rather than rational. However, our timing can be improved through intentionally noticing the flow of key events trends.

What we know for sure is that the world unfolding around us is too complex to fully understand all of the forces and factors involved. We cannot figure out in our rational minds how to connect all the dots. We often can't make sense of all of the moving parts in our world and so just declare it nonsense. However, watching as the dots emerge provides the basis for making it make sense.

There are indicators we can learn to look for; what we should notice. Watching something unfold is the key to good timing. We can learn what factors are in the critical path and identify key indicators that provide vision into the events, trends, or cycles as they unfold. But, we also have to allow our intuitive sense of the flow to be integrated in how we make choices.

There are indicators we can learn to look for; what we should notice.

After Mark's team had begun discussing their parallel development cycles, it was clear there were some important differences and a few critical conflicts. When I further examined their projections of timing, it was clear that the engineering group had a detailed work plan and estimates, while the marketing group had only a vague sense of how long different phases of market development would take.

Some of that is natural between these two functional groups. But in this case it was more. In fact, the senior marketing person was very experienced and relied upon his gut to make guesstimates. On the other hand, the engineering team was assuming significant time-savings in their technology development areas based upon recent automation. Both made good cases and we all learned from their patient walk-through of their timelines.

Opportunity Vision

An opportunity is an interpretation of the environment which contains the potential for future benefit. An opportunity vision has an inherent time frame for when the future benefit will occur.

Projecting expected cycle time and potential future benefit is based upon a person's subjective frame of reference. In fact, almost every word in the definition above has a subjective basis. The key to the subjective nature of the concept is "interpretation." No situation is inherently an opportunity; it must be defined as an opportunity based upon someone's interpretation; which is the process of applying one's own memories, perceptions, cognitive biases, education, financial preferences, intentions, values, assumptions, beliefs, etc.

Projecting expected cycle time and potential future benefit is based upon a person's subjective frame of reference.

Determination of the relevant environmental elements is also a subjective process. What factors are important, what resources are available, what future events will occur, what time frame is assumed, and many other questions are answered based upon the point of view of the observer. In addition, many of the factors of the environment are not consciously considered by a leader but are submerged in his or her subjective map of reality, or worldview.

A wide range of uncertainty is contained in the concept of "potential." The probability of reaching the desired destination depends upon numerous factors. The notion of "future benefit" is also highly subjective. A strategy of rapid growth fueled by leverage may be viewed by one individual as having an attractive future value, while another views the future value of a less-risky strategy of steady cash flow, low debt, and security as less attractive.

The subjective nature of an opportunity emphasizes another important point—a leader's perception of an "opportunity" may be inaccurate. The accuracy of his or her assessment of the environment largely depends on the many subjective assumptions incorporated into his or her model of reality. Significant differences may exist between the actual potential of a specific environment and the perceived potential of that environment as an opportunity. Given the time frames of most business opportunities, only the passage of considerable time will reveal erroneous gaps in a leader's perception.

I have taught Entrepreneurship at multiple major universities. I define "entrepreneurship" as recognizing change as opportunity and developing an effective organization response to that opportunity vision. An opportunity vision is clarity about the path to the innovation required to match changing circumstances. I have often opened speeches about innovation and entrepreneurship with the quote, "An entrepreneur is someone who shoots for an opportunity no one else sees. A successful entrepreneur is one that hits it."

All entrepreneurs, and good senior leaders, are driven by what they perceive as their opportunity vision. It is always out in time and over shifting terrain, but it provides the basis for their strategy and eventual success.

Expectation/Observation/Adaptation

This concept is captured in another triplet. The triplet discussed reflects logical steppingstones to better timing.

Expectations always include a time horizon. Whether our expectations concern ourselves, those around us, and the world we live in. We pay attention to things when we have attached expectations to them. Clarity about the expected timing of future results is important for self and others.

Observing a process in action allows a person to monitor key variables and re-evaluate timing expectations. Observation is based upon knowing what to notice and creating feedback loops that provide access to the key data points that need to be taken into consideration.

Adapting our mental algorithms is based upon integrating new information, experiences, and learning. When things change, our mental maps of reality need to change. In some cases, disruptive events seemed random, but in other cases there was a change in the system that needed to be acknowledged by adapting our mental maps.

We get frustrated when things take longer than our anticipated timing. Instead of recognizing that it is our mental algorithms that need to be adapted. We cannot change the past, but we can improve our timing skills so that we do not make the same mistake next time.

Feedback loops provide the opportunity to adapt our map of reality. Instead of being frustrated when something takes longer than you expected it would, it is much better to adapt our mental algorithms and maps we use to estimate timing rather than waste time blaming reality for not meeting your expectations.

Catching up with Mark and his team almost a year later, I was pleased to get a good report. Mark was pleased to share that the business process diagrams we had developed together had been the key to their high-speed performance. He said that by having established agreed-upon key dates, it had pushed all three teams to drive for those critical time windows.

By having established clear expectations and knowing what to monitor, they had been able to quickly adapt to changing pushes and pulls. It turned out that none of the teams were correct in their original forecasts because they all consistently finished ahead of the critical path requirement. By being able to rapidly adapt, they had out-performed their own expectations.

Leadership Practice: Noticing others' sense of timing

Description:

Again, these key practices are intended to stretch your mental habits and get you to notice and respond to different sensory input. In this case, the practice is focused on developing the ability to assess someone's sense of timing and pay attention to how they think. Not necessarily *what* they think, but it's more about observing *how* someone thinks. This insight can then be translated back into your own thought programs.

Process Steps:

Step 1 – Think through the concept of someone "having good timing." What does that mean? How does one distinguish between "good timing" and "not-so-good timing"?

Step 2 – Set a trigger to remind you to notice other people's "sense of timing" meetings. You will find most people tend to be overly optimistic or pessimistic about how long projects will take.

Step 3 – For those people that you think demonstrate good timing, ask them how they came up with their time estimates. They often cannot express their mental algorithms, but they can give you a sense of how they think about timing.

Step 4 – Practice using different ways described by others of noticing and estimating when key events and trend changes will occur.

Practice Opportunity Triggers:

A good place to explore this practice is in team meetings. When deadlines are being discussed, use that as a reminder, or trigger, to explore different people's sense of timing on key events. Notice over time who tends to have a good time estimator.

TQ Organizational Development Implications

Leadership Triads for Alignment

One of the key implications in Jaques's organizational theory is the need for a leader to regularly communicate with both his or her direct reports and the direct reports 'once removed'. There are numerous benefits to having structured interaction between three levels of the organization. One of the most important is to align around time frames.

According to Jaques, each organizational level should be managing a different time frame. Regular communication can help a manager understand the operating time frames and priority cycles for both of their direct reports and also their subordinates once-removed. All three levels are working together to integrate shorter-term cycles into medium-term cycles into the timespan of the manager's role.

Jaques proposes the following:

A little-recognized fact in organizations is that the relationship between managers once removed (MoRs) and their subordinates once removed (SORs) is critical to the well-being of every managerial accountability hierarchy and its future strengths.

The priority role responsibilities are different at the different levels of the organization, but the lower levels are nested as a part of the higher level.

The problems being solved at the lower levels have a shorter time horizon but account for critical cycles for completion of the higher-level processes. By having regular structured triad leadership triads, important insight can be gained by all three levels' effective alignment.

Reflections and Action

- Back to the beginning of the chapter, how good is your sense of timing? A first step is to begin to notice your own and other's ability to estimate cycle time. What can you do to begin noticing this subtle, but important, TQ capacity?

- Think about the following statement, "Our sense of how things will evolve is essentially intuitive rather than rational. However, our timing can be improved through intentionally noticing the flow of events and trends and practicing estimating cycle time." Do you agree with this statement? Why does it matter whether timing is intuitive our rational? The key point is that both ways of knowing can be improved with intentional practice.

- Another direct quote from the chapter, "We are multi-sensory beings with a wide range of "sensing". We are often confused that our five primary external senses are our only way of making sense out of our world. Just think about our sense of balance, our sense of humor, our sense of danger, our sense of right and wrong. One of our key senses is our sense of timing." Have you thought about the fact that we are not limited to just the five primary senses but have multiple senses that help us make "sense of the world"? What are some of the other senses that play an important role in your leadership style?

- Have you heard the expression, "You will not get what you 'expect', but what you 'inspect'?" There is a degree of truth to this old expression. Simply having an expectation of yourself or someone else does not mean it will get done. However, when we communicate our expectations and then follow up to monitor progress, we are more likely to get what we expected and better able to adapt if not. What are some of the key things you monitor in your leadership role to assure plans are on track?

CHAPTER TEN

Improving Timing Skills

Is timing innate or can it be developed? Well, some of both. We have all developed our sense of timing as a synthesis of learning from experience and an openness to our intuition. But like learning to play a musical instrument, some people seem to have a natural aptitude, while others struggle with basic rhythm. But in either case, everyone can get better with practice.

The previous two chapters discussed how important it is for a good leader to understand the cycles they manage and to effectively anticipate the timing of the key events and trends within cycles. This chapter is focused on how we can develop our sense of timing and timing skills.

Thesis: The thesis of this chapter is that everyone can improve their sense of timing by intentionally paying attention to cycle time and cycle-timing. Based upon noticing how long different processes and cycles take to complete, we can adapt our mental algorithms and maps.

Everyone can improve their sense of timing by intentionally paying attention to cycle time and cycle-timing.

There are a variety of ways that we use different timing skills when making decisions and estimating outcomes. In general, short-cycle processes are easier to predict. For example, estimating when a sale will close is different than estimating the time required for a merger to be completed. Which is still much simpler than estimating when a new technology will be developed and deployed.

Independent of the timing of the cycle of interest, we all can get better at estimating when a process will produce its outcome, and then begin again. Nevertheless, working with trends, processes and cycles that take years is a greater challenge than working with shorter life cycles.

For one reason, the opportunity for feedback is stretched out much further. Feedback is essential for adapting to changing circumstances throughout the cycle. In many cases, it takes years before we can determine the accuracy of our predictions. The mandate of a leader is that they must regularly make decisions based upon limited information.

Timing is important in a whole range of areas. Consider timing in a relationship, timing in organizational changes, timing in market shifts. In many areas of leadership, good timing is an essential capacity elite performer.

TQ Concepts to be discussed in this chapter:

- Sorting Programs
- Sorting for Cycles by Time Frames
- Frustration as a sign of poor timing

Leadership Practice: Refining Timing Algorithms

• • • • •

Case Study Scenario

Listening to Mark explain how his team had exceeded his expectations, it was clear to me that he was describing how his team had upgraded their timing skills. Building rapid detection and adaptation loops into their operating procedures had paid immediate dividends.

We discussed that for a few minutes, and he asked if I would come back in and do a "lessons learned" workshop, and I agreed. This opened the door to explore a new set of TQ concepts. And as it turned out, help Mark build better timing algorithms into their ongoing processes.

The first concept we discussed was "sorting programs".

TQ Concepts

Sorting Programs

Sorting programs are powerful filters that determine what you notice. When washing clothes, for example, you sort by whites and darks. When you walk into a crowded room, you sort for someone you know. When you walk into a parking lot, you sort for your car. You are always sorting through the virtually infinite number of things that you could be paying attention to. You sort to help you focus on specific things. "Sort programs" determine what you will perceive and think about. Typically, we sort according to patterns developed unconsciously. Most of the time, we are unaware of what we sort for.

Sorting programs are powerful filters that determine what you notice.

It is important to note that sorting programs can be analyzed and modified. For instance, when I am in a new situation, I choose to sort for "delightful." I try to notice things that delight me—and I usually find them. On the contrary, I have known many people that sort for something to complain about—and they usually find them too. Developing the ability to adapt your sorting programs gives you a great deal of control over what you experience and how things unfold.

In Mark's case, what they were sorting for were different things in different departments. In the technology development group, they were sorting for MVP completion at the module level. Marketing was sorting for early adopters and potential competitors. Production was sorting for supply bottlenecks. And finance, they always sort for actual vs. budget.

Sorting for Cycles by Time Frames

One of the sorting filters we can use is to sort by time frame. This involves setting triggers to notice the timing of things as they unfold. This allows us to adapt our expectations about the expected timing of the outcome. The objective is to sort for relevant trends that occur with different time frames. This helps us notice the time intervals of nested cycles.

So, what are the cycles and trends relevant to your leadership role? At first, most people have a hard time identifying what is relevant in the flow of processes in their organization, their industry, and the broader world. However, everybody has a set of activities they are focused on at any moment. The key is to identify the broader cycles we are nested within so that we can keep track of them.

When I was a marketing VP with a large IT systems integrator, I had decided to switch from a weekly activity report to a weekly "mistake report". Most of us spend much of our time trying to avoid and/or deny mistakes. I wanted them sorting for mistakes. I never criticized them for the mistake they reported but would praise them for recognizing it—and admitting it.

I defined a "mistake" as something you would do differently next time. The only way we can get better is to change and noticing what needs to change is sometimes the hardest part. By sorting for new mistakes, I helped them identify a lot of little places where they could change to get better.

I did have one catch. They could only use a mistake once. The problem is often that we make the same mistake over and over again. Sometimes even when we recognize the pattern. My only one credit for each new mistake became the driver. It caused my team to constantly sort for opportunities for improvement because that is what they were sorting for. I had a sign on my desk that said, "Always make new mistakes".

Frustration as an emotional signal of poor timing

When our estimation of timing is wrong, the feeling we experience is often frustration. We can use our frustration as a trigger to adapt our mental algorithms. We tend to confuse our estimation as when something "should" have occurred.

Frustration is defined as the emotion experienced when one's expectations have not been met by reality. When we feel frustrated, we are often dealing with our expectations around timing. This emotion can signal you to think about why you estimated poorly and learn to adapt your mental algorithms and not just get upset with reality.

Frustration is defined as the emotion experienced when one's expectations have not been met by reality

---

I used to get extremely frustrated in traffic. A trip that I expected to take twenty minutes would end up taking an hour, and I would experience the emotion of frustration. I do not like feeling frustrated; most of us don't. Yet, when I set out on a drive expecting it to take an hour, I am not frustrated when it does take an hour. Frustration is experienced when expectations do not match actual circumstances. A high percentage of our frustration comes from expectations around "when" something will happen.

Leadership Practice: Refining Timing Algorithms

Description:

In the previous chapter, we discussed mental algorithms that are part of how we estimate cycle time. Here we're looking at mental algorithms from a development perspective. How can we consciously get insight into our subconscious algorithms to adapt them to be more effective based on changing circumstances.

When we overestimate or underestimate a time frame, we should ask ourselves, Why did I think it would happen earlier?" What was my reasoning behind the timing of my expectation and where did it go wrong? Developing our timing algorithms is based on learning from experience and adapting to feedback.

Process Steps:

Step 1 – The first requirement is to create a meta-perspective that allows you to reflect on your own thought processes. We need insight into our own operating thought patterns to be able to adapt them to reflect experience.

Step 2 – Intentionally pay attention to how your sense of timing works. Notice when you are surprised by the timing of something.

Step 3 – Practice reflecting on your timing algorithms to evaluate what works and what does not. The more we understand how we subconsciously estimate time, the more we can adapt our mental models and develop better time-estimation skills.

Practice Opportunity Triggers:

Refining our mental algorithms to improve our sense of timing can be triggered when our time estimates are wrong. The functioning of our timing algorithms is most apparent when we realize our estimation was inaccurate. These opportunities can serve as a trigger to allow you to understand where your estimator overestimated or underestimated how long things would take.

Frustration can be a great trigger to reflect on your sense of timing. Noticing when we are feeling frustrated can be used as a useful trigger for reflecting on your miscalculated time expectations.

TQ Organizational Development Implications

Three Versions of Organizational Structure

Improving cycle-timing is essential about better understanding the cycles within cycles that lead to outcomes in different cycles. It seems it would be simple enough to diagram the key business processes and life cycle dynamics, but it turns out the complexity is too overwhelming. In addition, which cycles we focus on depends upon our role and time frame of responsibility.

Another complicating factor is that there are always three different versions of the "organization." There is the official version on organization charts; there is the way people might think the organization works; and the actual activities and decisions being made.

Jaques describes the three versions of the organization as:

***Manifest organization**: the organization structure as it is represented on the official organization chart; at best, only a very rough approximation to what is actually going on, if you can even make sense of it.*

***Assumed organization**: the structure is different; people assume it really works; is likely to have as many variants as you have people. And produces confusion.*

***Extant organization**: the system as it actually functions. It can be demonstrated with systemic study; the model will always be an approximate picture. It requires that you dig in and find who is actually being held accountable for what and what authority they are in fact able to exercise in relationship to whom and to what.*

Developing better cycle-timing is a function of understanding the Extant Organization and the interdependent dynamics among its subsystems. An important insight is that we all, each of us, is using an "assumed version" of how the organization works. Much of our frustration comes from the reality that our mental model is wrong.

We operate in a sea of assumptions. When our assumptions are not aligned with "extant version of how things are organized," or how things actually happen. Dynamics can never adequately be modeled but it is important to keep in mind that "how things work" is always different than "how you think things work."

Reflections and Actions

- Do you feel like you have a handle on how to go about monitoring and adapting your mental time estimation programs? The concept is straightforward but putting it into practice can be challenging. Learning new programs is uncomfortable, but worth the effort. What is one specific action you can take to begin working with your predictive mental algorithms?

- One quote in the chapter worth reflecting upon is "Sort programs determine what you will perceive and think about. Typically, we sort according to patterns developed unconsciously. Most of the time, we are unaware of what we sort for." Are you familiar with the concept of sort patterns? It is a very valuable metaphor. How aware are you of your current default sort patterns around timing?

- What are the cycles and trends relevant to your leadership role? This is a very important question. You need to be paying attention to the right things. Every role has a set of priority areas of responsibility and expectations. These responsibilities and expectations contain a wide variety of factors over long periods of time. It is good practice to develop an inventory of the key cycles and trends relevant to your current leadership role.

- "Frustration is defined as "the emotion experienced when one's expectations have not been met by reality." When we feel frustrated, we are often dealing with our expectations around timing. This emotion can signal you to think about why you estimated poorly and learn to adapt your mental algorithms. How easily frustrated are you? Can you think of someone that seems to be constantly frustrated that things take too long? Reality has a way of unfolding that we need to pay attention to and not be criticizing.

• • • • •

Case Study Conclusion

Aligning the multiple cycles in the new opportunity life cycle is a constant challenge for leaders, especially in tech-driven sectors. Each part of the cycle is dependent upon the other parts—each with their own time frames. The art comes with how a leader balances the dependencies of each sub-cycle on the other cycles.

Back to our case study for this third part, let me just say, Mark struggled. Trying to get each of his departments to provide an estimated range for the amount of time their cycle would take was not easy. Getting them aligned and understanding the interdependencies was hard. Every group could provide a reasonable target for the time they would require but had difficulty actually staying committed to it.

In the end, Mark was able to get his team aligned and they pulled it off. They launched their next product suite just in time to catch the market wave they had forecasted. But it was not without a lot of conversation and negotiation. Each group had their own version of the required cycle time and the timing the investments required to support the strategy.

After a few months, Mark called and shared, *"I do not know whether to blame you or congratulate you. You helped us develop a new product suite with amazing teamwork by helping us come to grips with our interdependent cycles. The TQ concepts helped us to align internally and, as it turned out, we were in the perfect position to take advantage of our technical innovations in an uncertain market. Thank you."*

Always nice to hear.

And thank you for staying with us this long. We hope you can make these TQ concepts and practices as valuable as they are interesting. ■